FORTSCHRITTE DER PHYSIK

HERAUSGEGEBEN IM AUFTRAGE DER PHYSIKALISCHEN GESELLSCHAFT
DER
DEUTSCHEN DEMOKRATISCHEN REPUBLIK
VON F. KASCHLUHN, A. LÖSCHE, R. RITSCHL UND R. ROMPE

HEFT 2 · 1981 · BAND 29

AKADEMIE-VERLAG · BERLIN

EVP 10,– M
31728

ISSN 0015-8208

BEZUGSMÖGLICHKEITEN

Bestellungen sind zu richten

- **in der DDR** an eine Buchhandlung
  oder an den AKADEMIE-VERLAG,
  DDR - 1080 Berlin, Leipziger Straße 3—4
- **im sozialistischen Ausland** an eine Buchhandlung für fremdsprachige Literatur oder an den zuständigen Postzeitungsvertrieb
- **in der BRD und Berlin(West)** an eine Buchhandlung oder an die Auslieferungsstelle
  KUNST UND WISSEN, Erich Bieber OHG,
  7000 Stuttgart 1, Wilhelmstraße 4—6
- **in Österreich** an den Globus-Buchvertrieb,
  1201 Wien, Höchstädtplatz 3
- **in den übrigen westeuropäischen Ländern** an eine Buchhandlung oder an die Auslieferungsstelle
  KUNST UND WISSEN, Erich Bieber GmbH,
  CH - 8008 Zürich/Schweiz, Dufourstraße 51
- **im übrigen Ausland** an den Internationalen Buch- und Zeitschriftenhandel; den Buchexport, Volkseigener Außenhandelsbetrieb der Deutschen Demokratischen Republik,
  DDR - 7010 Leipzig, Postfach 160; oder

  an den AKADEMIE-VERLAG,
  DDR - 1080 Berlin, Leipziger Straße 3—4

Zeitschrift „Fortschritte der Physik“

Herausgeber: Prof. Dr. Frank Kaschluhn, Prof. Dr. Artur Lösche, Prof. Dr. Rudolf Ritschl, Prof. Dr. Robert Rompe, im Auftrag der Physikalischen Gesellschaft der Deutschen Demokratischen Republik.
Verlag: Akademie-Verlag, DDR - 1080 Berlin, Leipziger Straße 3—4; Fernruf: 2236221 und 2236229; Telex-Nr. 114420; Bank: Staatsbank der DDR, Berlin, Konto-Nr. 6836-26-20712.
Chefredakteur: Dr. Lutz Rothkirch.
Anschrift der Redaktion: Sektion Physik der Humboldt-Universität zu Berlin, DDR - 1040 Berlin, Hessische Straße 2.
Veröffentlicht unter der Lizenznummer 1324 des Presseamtes beim Vorsitzenden des Ministerrates der Deutschen Demokratischen Republik.
Gesamtherstellung: VEB Druckhaus „Maxim Gorki“, DDR - 7400 Altenburg, Carl-von-Ossietzky-Straße 30/31.
Erscheinungsweise: Die Zeitschrift „Fortschritte der Physik“ erscheint monatlich. Die 12 Hefte eines Jahres bilden einen Band.
Bezugspreis je Band 180,— M zuzüglich Versandspesen (Preis für die DDR: 120,— M). Preis je Heft 15,— M (Preis für die DDR: 10,— M).
Bestellnummer dieses Heftes: 1027/29/2.

AN (EDV) 57618

Fortschritte der Physik 29, 35–94 (1981)

Meson Interactions within Nonlinear Chiral Theories

D. EBERT[1]) and M. K. VOLKOV

Joint Institute for Nuclear Research, Dubna, USSR

Contents

[1]) On leave from Institut für Hochenergiephysik, Berlin-Zeuthen, DDR.

I. Introduction

It has become widely accepted that strong interactions exhibit a set of approximate symmetries corresponding to the chiral groups $SU(2) \times SU(2)$, $SU(3) \times SU(3)$ and, perhaps, $SU(4) \times SU(4)$. These groups are generated by the algebra of vector and axial-vector currents of hadrons. The requirement of an approximate chiral symmetry of strong interactions, realized in the Lagrangian formalism allows, in particular, a deeper understanding of the successful results of current algebra, low-energy theorems and the PCAC principle (partial conservation of the axial vector current) [*1*, *2*].
Chiral symmetry $SU(N) \times SU(N)$ $(N = 2, 3, 4, \ldots)$ differs from the usual $SU(N)$ symmetry of hadrons by the fact that, in order to be applicable to strong interactions, it must be spontaneously broken by the vacuum. It gives no rise therefore to new conserved physical quantities and particle classification. Clearly, this situation is opposite to the case of a theory of free massless spinor particles, e.g. neutrinos, where chiral symmetry implies helicity conservation. In a world with massive baryons, chiral symmetry can be restored only by introducing additional pseudoscalar massless particles (Goldstone bosons). Instead of new conservation laws, it leads here to restrictions on the possible shapes of the interaction of the baryons with the Goldstone bosons. Hence, chiral symmetry is often called a dynamical symmetry.
In a chiral-symmetric theory baryon multiplets are degenerated in mass, and the analogs of the π-, K- and η-mesons are massless. Since in reality we are concerned with baryons of different masses and massive mesons, the spontaneous breakdown of vacuum symmetry must be further supplemented by a suitable mechanism of explicit symmetry breaking.
Historically the first attempts to construct a chiral-symmetric field theory on the basis of the Lagrange formalism were done by SCHWINGER [*3a*] and GELL-MANN and LEVY [*4*]. They proposed the so-called σ-model the linear version of which gives rise to a renormalizable chiral-invariant field theory. On the other hand, there exists also a nonlinear version of this model without fictious σ particles which is nonrenormalizable [*3b*, *4*]. Furthermore, nonpolynomial chiral field theories have been devised independently by GÜRSEY [*5*]. Lateron it was shown by WEINBERG [*6*] that the linear σ-model turns into a nonlinear one if the mass of the σ particle tends to infinity. The most general technique of constructing nonlinear chiral Lagrangians was developed in works by COLEMANN, WESS and ZUMINO [*7*] and D. V. VOLKOV [*8*].
Despite of their nonrenormalizability nonlinear chiral Lagrangians have certain advantages as compared to linear theories. First, as has been mentioned, nonlinear models are free of unphysical σ particles (note that a linear $SU(3) \times SU(3)$ $(SU(4) \times SU(4))$ σ-model contains 8 (15) such particles). Second, even in the "tree" approximation nonlinear chiral theories reproduce quite simply all the results of low-energy theorems. The advantages of using nonlinear chiral Lagrangians become especially evident in considering multi-meson processes where the use of standard reduction and soft-pion techniques is very combersome. The Lagrangian formalism allows one also to make the next step in perturbation theory and to consider the one-loop approximation that contains essentially new information as compared to the simplest tree approximation. This point of view has been, in particular, emphasized in the works of ref. [*9*].
This review provides an introduction to the field of nonlinear chiral Lagrangians and their possible physical applications and is intended for readers slightly acquainted with

that domain of quantum field theory. The whole content of the review can be splitted into two parts. The first of them (Sect. II and III) is of a rather formal character. It contains, for the sake of clarity and in order to make the paper relatively self-contained, the principle theoretical ideas behind the Lagrangian approach. In particular, the structure of nonlinear $SU(N) \times SU(N)$ ($N = 2, 3, 4, \ldots$) meson-baryon and meson-quark Lagrangians is discussed in detail. Particular attention is given to the construction of unified gauge models for the weak and electromagnetic interactions of hadrons. Specifically, for the gauge group $SU(2)_L \times U(1)$ a nonlinear model of the Weinberg-Salam-type results, which is expressed directly in terms of observable hadrons.
The second part of the review (Sect. IV) is devoted to applications of chiral Lagrangians to low-energy interactions of pseudoscalar mesons. Special attention is paid here to the one-loop approximation that gives not only corrections to the Born (tree) approximation but provides also, as mentioned above, essentially new physical information. We calculate here such physical quantities as scattering lengths, form factors, meson polarizabilities, decay amplitudes and structure constants. The results of calculations are in good agreement with experiment for the groups $SU(2) \times SU(2)$ and $SU(3) \times SU(3)$.

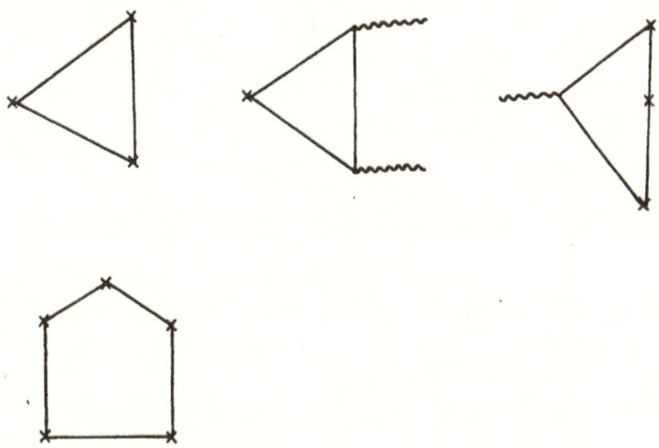

Fig. 1. Various types of anomalous baryon loop diagrams. • — vector vertices, × — axial vector vertices

It is worth remarking that many of the decay processes considered proceed via so-called anomalous loop diagrams [*10*]. The most important anomalous baryon loop diagrams containing axial-vector and vector vertices are shown in Fig. 1. For the heavier charmed mesons F, D and η_c chiral symmetry is expected to be very approximate. Decay processes with charmed mesons have been studied therefore only in the tree approximation.
As to the regularization of divergent meson loop diagrams, special (superpropagator) methods applicable to theories with nonpolynomial Lagrangians have been used. On the other hand, the one-loop baryon diagrams may be regularized by standard methods of renormalizable field theory.
Sect. II introduces the technique of nonlinear Lagrangians. Sect. III is devoted to the construction of unified gauge models based on chiral Lagrangians, to the choice of a proper mechanism of symmetry-breaking and to a study of the role of the Cabibbo angle. Applications of a chiral meson-baryon Lagrangian are presented in Sect. IV and compared with analogous results obtained within a meson-quark Lagrangian. Discussions and a short outlook are given in Sect. V.

II. A Survey of Nonlinear Chiral Lagrangians

II.1. Chiral $SU(2) \times SU(2)$ σ-model

It is instructive to examine field theory models which explicitly show invariance under the chiral group. The best known Lagrangian model based on $SU(2) \times SU(2)$ is the so-called σ-model [3, 4]. There exist also further extensions of the model to the higher groups $SU(3) \times SU(3)$ [11] and $SU(4) \times SU(4)$ [12]. The σ-model has the attractive feature of vacuum symmetry breaking thereby allowing for the generation of fermion masses. Furthermore, as has been shown by WEINBERG [6] and BARDEEN and LEE [13], the linear σ-model converts into a nonlinear one if the mass of the σ particle is allowed to become very large. The nonlinear realization of chiral $SU(2) \times SU(2)$ was one of the starting-points for further generalizations leading to phenomenological chiral Lagrangians of higher chiral groups [7, 8] (for earlier works on nonlinear phenomenological Lagrangians see also ref. [5, 14]). It should be noticed that the nonlinear chiral Lagrangians are non-renormalizable in the usual sense, but they have the remarkable property of reproducing the soft-pion theorems of current algebra [6, 15] in the tree-approximation.
Let us now briefly discuss the linear and nonlinear versions of the σ-model in order to introduce the reader into the basic ideas of nonlinear realizations of symmetry groups. The Lagrangian of the renormalizable linear σ-model is

$$\mathcal{L} = \frac{1}{2}\left[(\partial_\mu \boldsymbol{\pi})^2 + (\partial_\mu \sigma)^2 - \mu^2(\boldsymbol{\pi}^2 + \sigma^2)\right] + \frac{\lambda}{4}(\boldsymbol{\pi}^2 + \sigma^2)^2$$
$$+ i\bar{N}\gamma_\mu\,\partial^\mu N + g_{\pi N}\bar{N}(\sigma + i\gamma_5\boldsymbol{\tau\pi})\,N, \qquad \left(\gamma_5 = -\begin{pmatrix} 0 & I \\ I & 0 \end{pmatrix}\right) \tag{2.1}$$

where $N = \begin{pmatrix} p \\ n \end{pmatrix}$ is a nucleon doublet field, $\boldsymbol{\pi}$ is the isospin-one pion field and σ is an isoscalar scalar field; τ_i are usual Pauli matrices. It is convenient to rewrite the Lagrangian (2.1) in terms of the meson matrix Π,

$$\Pi = (\sigma 1 + i\gamma_5 \tau \cdot \pi), \qquad \left(\tau \cdot \pi \equiv \boldsymbol{\tau\pi} = \sum_{i=1}^{3} \tau_i \pi_i\right),$$

as

$$\mathcal{L} = \frac{1}{16}\,\mathrm{Tr}\,[\partial_\mu \Pi^+\,\partial^\mu \Pi - \mu^2 \Pi^+ \Pi] + \frac{\lambda}{32}\,\mathrm{Tr}\,(\Pi^+\Pi)^2 + i\bar{N}\gamma_\mu\,\partial^\mu N + g_{\pi N}\bar{N}\Pi N, \tag{2.2}$$

where the trace is taken over internal and Lorentz indices.
Let $V_i = \tau_i/2\,1$ and $A_i = \tau_i/2\,\gamma_5$ ($i = 1, \ldots, 3$) denote the matrix representations of the generators of isospin and isoaxial transformations of chiral $SU(2) \times SU(2)$ which satisfy the commutation relations and normalization conditions[2])

$$[V_i, V_j] = i\varepsilon_{ijk}V_k, \quad [V_i, A_j] = i\varepsilon_{ijk}A_k, \quad [A_i, A_j] = i\varepsilon_{ijk}V_k;$$
$$\mathrm{Tr}\,A_iA_k = \mathrm{Tr}\,V_iV_k = 2\delta_{ik}, \quad \mathrm{Tr}\,A_iV_k = 0, \tag{2.3}$$

[2]) Forming the "left" and "right" combinations $V_i^\pm = V_i \pm A_i$, we get $[V_i^\pm, V_j^\pm] = i\varepsilon_{ijk}V_k^\pm$, $[V_i^+, V_j^-] = 0$, and (2.3) factors into two commuting $SU(2)$.

ε_{ijk}, δ_{ik} being the antisymmetric tensor and Kronecker delta symbol, respectively. It then follows that the Lagrangian (2.2) is invariant under the isospin transformations

$$N \to N' = e^{i\beta\cdot V} N, \qquad (\beta \cdot V = \beta_i V_i)$$
$$\Pi \to \Pi' = e^{i\beta\cdot V} \Pi e^{-i\beta\cdot V} \tag{2.4}$$

and under chiral transformations[3])

$$N \to N' = e^{i\alpha\cdot A} N$$
$$\Pi \to \Pi' = e^{-i\alpha\cdot A} \Pi e^{-i\alpha\cdot A}. \tag{2.5}$$

The infinitesimal version of the transformation laws (2.4), (2.5) is

$$\delta\Pi = \beta_i \delta_i \Pi, \qquad \delta^5\Pi = \alpha_i \delta_i^{\,5} \Pi$$

or

$$\delta_l \pi_r = \varepsilon_{lri} \pi_i, \qquad \delta_l \sigma = 0 \tag{2.4'}$$

$$\delta_l^{\,5} \pi_r = -\delta_{lr} \sigma, \qquad \delta_l^{\,5} \sigma = \pi_l. \tag{2.5'}$$

Let us now consider the vector and axial vector currents $J_{i\mu}$, $J^5_{i\mu}$ and the associated charge operators Q_i and $Q_i^{\,5}$ of the model. By standard methods one gets (cf. Sect. III 2.2)

$$J_{i\mu} = -\frac{\delta \mathcal{L}}{\delta(\partial^\mu \varphi_r)} \delta_i \varphi_r, \qquad J^5_{i\mu} = -\frac{\delta \mathcal{L}}{\delta(\partial^\mu \varphi_r)} \delta_i^{\,5} \varphi_r, \tag{2.6}$$

φ_r being the respective hadron fields. The charge operators Q_i and $Q_i^{\,5}$ are just the generators of the chiral group acting in the Hilbert space of the fields. They are defined by

$$Q_i = \int J_{i0}(x)\, d^3x = -\int P_r \delta_i \varphi_r(x)\, d^3x$$

with an analogous expression for $Q_i^{\,5}$; $P_r = \partial\mathcal{L}/\partial(\partial_0 \varphi_r)$ is the canonical momentum. Using the canonical commutation relations

$$[\varphi_r(\boldsymbol{x}, t), P_s(\boldsymbol{y}, t)] = i\delta_{rs} \delta^{(3)}(\boldsymbol{x} - \boldsymbol{y}),$$

one obtains

$$[Q_i, \varphi_r(\boldsymbol{x}, t)] = i\delta_i \varphi_r(\boldsymbol{x}, t) \quad \text{etc.}$$

The last equation together with eqs. (2.4'), (2.5') yields the set of commutation relations

$$[Q_i, \pi_j] = i\varepsilon_{ijk} \pi_k, \qquad [Q_i, \sigma] = 0,$$
$$[Q_i^{\,5}, \pi_j] = -i\delta_{ij} \sigma, \qquad [Q_i^{\,5}, \sigma] = i\pi_i, \tag{2.7}$$

which is the operator version of chiral field transformations. Finally, let us give the explicit expressions of the currents

$$J_{i\mu} \equiv V_{i\mu} = \overline{N} \gamma_\mu V_i N + (\boldsymbol{\pi} \times \partial_\mu \boldsymbol{\pi})_i,$$
$$J^5_{i\mu} \equiv A_{i\mu} = \overline{N} \gamma_\mu A_i N + (\sigma \partial_\mu \boldsymbol{\pi} - \boldsymbol{\pi}\, \partial_\mu \sigma)_i. \tag{2.8}$$

---

[3]) A transformation law of the form (2.5) for the Π-field is suggested by requiring invariance of the Yukawa coupling $\overline{N}\Pi N$ under chiral transformations of N. A more formal argument can be found in the text following eq. (2.25).

Till now the nucleons appearing in the Lagrangian (2.1) are massless. In the case $\mu^2 < 0$, $\lambda < 0$ the model exhibits the attractive possibility of getting massive nucleons due to a nonvanishing vacuum expectation value $\langle\sigma\rangle_0 \neq 0$ of the σ field. Indeed, seeking for the ground state of the model for constant field values by taking $\delta\mathcal{L}/\delta\sigma = 0$, $\delta\mathcal{L}/\delta\pi_i = 0$, one easily gets

$$\langle\sigma\rangle_0 = \pm\left(\frac{\mu^2}{\lambda}\right)^{1/2}, \qquad \langle\pi_i\rangle_0 = 0. \tag{2.9}$$

Let us choose $\langle\sigma\rangle_0 = -(\mu^2/\lambda)^{1/2} \equiv -F$ and set $\sigma = \sigma' - F$ where now $\langle\sigma'\rangle_0 = 0$. In terms of the new field σ' the Lagrangian (2.1) reads

$$\mathcal{L}(\sigma') = \frac{1}{2}\,[(\partial_\mu\boldsymbol{\pi})^2 + (\partial_\mu\sigma')^2] - \frac{1}{2}\,(-2\mu^2)\,\sigma'^2 + \frac{\lambda}{4}\,(\boldsymbol{\pi}^2 + \sigma'^2)^2 - F\lambda\sigma'(\boldsymbol{\pi}^2 + \sigma'^2)$$
$$+ \overline{N}(i\gamma_\mu\,\partial^\mu - M)\,N + g_{\pi N}\overline{N}(\sigma' + i\gamma_5\tau\cdot\pi)\,N, \tag{2.10}$$

apart from an unimportant constant, where M is the nucleon mass, $M = g_{\pi N}F$. Besides nucleon mass generation, one gets also $\sigma\pi$-mass splitting, absent in the original Lagrangian (2.1). Evidently, by inspection of (2.10) we see that the pions are massless, while $m_{\sigma'}^2 = -2\mu^2 > 0$. This mass splitting signals spontaneous breakdown of chiral $SU(2) \times SU(2)$ symmetry due to vacuum noninvariance. Indeed, assuming $\langle\sigma\rangle_0 \neq 0$ we get from (2.7)

$$\langle 0|\,[Q_i^5, \pi_j]\,|0\rangle = -i\delta_{ij}\langle\sigma\rangle_0 \neq 0 \tag{2.11}$$

and thus

$$Q_i^5|0\rangle \neq 0. \tag{2.12}$$

It has been proved on general grounds of quantum field theory that spontaneous symmetry breakdown is accompanied by the appearance of massless particles, the Goldstone bosons [*16*]. In the σ-model these are just the pions. We have here the interesting case that the Lagrangian (2.1) is invariant under the group $SU(2) \times SU(2)$ generated by the charges Q_i, Q_i^5 while the vacuum is not. Such a situation is generally called Nambu-Goldstone realization of chiral $SU(2) \times SU(2)$, and there arise no implications for the classification of states with respect to this group.
Let us rewrite (2.10) in matrix notation. It is convenient to consider a canonical field transformation $(\sigma', \boldsymbol{\pi}) \to (\varphi, \Phi/F)$ defined by

$$\Pi = \big((\sigma' - F)\,1 + i\gamma_5\tau\cdot\pi\big) = (\varphi - F)\,e^{-i\Phi\cdot\tau\gamma_5/F}. \tag{2.13}$$

Then the Lagrangian (2.10) is

$$\mathcal{L} = \frac{1}{16}\,\mathrm{Tr}\,(\partial_\mu\Pi^+\partial^\mu\Pi) - \frac{\mu^2}{2}\,(\varphi - F)^2 + \frac{\lambda}{4}\,(\varphi - F)^4 + i\overline{N}\gamma_\mu\partial^\mu N$$
$$+ g_{\pi N}(\varphi - F)\,\overline{N}\exp\,(-i\Phi\cdot\tau\gamma_5/F)\,N. \tag{2.14}$$

The transition from the linear σ-model (2.14) to its nonlinear version can be simply achieved by considering the limit $(\mu^2, \lambda) \to -\infty$, $(\mu^2/\lambda) = F^2$ fixed [*6, 13*]. Since the mass term of the σ particle ($\sim -2\mu^2$) goes to infinity, reference to the scalar field φ (or σ') may be dropped. Then, we have

$$\Pi = -Fe^{-i2\xi\cdot A}, \qquad (\xi_i = \Phi_i/F) \tag{2.15}$$

and

$$\mathcal{L} = \frac{F^2}{16}\,\mathrm{Tr}\,(\partial_\mu e^{i2\xi\cdot A})\,(\partial^\mu e^{-i2\xi\cdot A}) + \overline{N}(i\gamma_\mu\partial^\mu - Me^{-i2\xi\cdot A})\,N. \tag{2.16}$$

The Lagrangian (2.16) defines the nonlinear σ-model in the socalled exponential parametrization frequently used in the literature [5, 17]. Re-expressing (2.15) in terms of the original fields σ, $\boldsymbol{\pi}$ we obtain now

$$\Pi\Pi^+ = F^2 = \sigma^2 + \boldsymbol{\pi}^2$$

showing that in the nonlinear σ-model the σ field is no longer independent. From the last formula we see that the isotopic space of meson fields becomes a curved one (a 3-dimensional sphere of radius F in a 4-dimensional space). Different forms of chiral Lagrangians [5] then correspond to different coordinate systems on this sphere [8]. In particular, the exponential parametrization used above is related to normal coordinates. To the first order in the parameter $1/F$ the Lagrangian (2.16) takes the form

$$\mathcal{L} = \frac{1}{2}(\partial_\mu \boldsymbol{\Phi})^2 + \overline{N}(i\gamma_\mu \partial^\mu - M) N + i g_{\pi N} \overline{N} \left[\gamma_5 \boldsymbol{\tau} \cdot \boldsymbol{\Phi} + O\left(\left(\frac{\Phi}{F}\right)^2\right)\right] N + O\left(\left(\frac{\Phi}{F}\right)^4\right). \tag{2.17}$$

Thus, to the lowest order in $1/F$ we get the usual pseudoscalar πN-Lagrangian, where $\boldsymbol{\Phi}$ is the new pion field. The remaining terms modify, however, the Lagrangian and destroy its renormalizability. As has been emphasized, the advantage of the nonlinear σ-model is that it retains the symmetry properties of the original model without introducing a hypothetical σ particle[4]). It should be noticed that one can further add a suitable symmetry breaking term $\Delta\mathcal{L}$ to the Lagrangian (2.1) to get a finite pion mass and a partially conserved axial vector current (PCAC), $\partial^\mu A_{i\mu} = F\mu^2\pi_i$.
Let us next rewrite the nonlinear Lagrangian (2.16) in a form which makes further generalizations possible and is closely related to the techniques of covariant derivatives considered in the subsequent Section. To this end, introduce a new nucleon field Ψ by

$$\Psi = e^{-i\xi\cdot A} N. \tag{2.18}$$

The Lagrangian (2.16) is then transformed into the equivalent expression

$$\mathcal{L}(\Psi, \xi) = \overline{\Psi} i\gamma^\mu \{\partial_\mu + [\exp(-i\xi\cdot A)\, \partial_\mu \exp(i\xi\cdot A)]\}\, \Psi - M\overline{\Psi}\Psi + \mathcal{L}_0(\xi), \tag{2.19}$$

$\mathcal{L}_0(\xi)$ being the meson Lagrangian. The differential expression in the square bracket of (2.19) may be expanded in terms of the complete set of generators V_i, A_i of chiral $SU(2)\times SU(2)$

$$\exp(-i\xi\cdot A)\, \partial_\mu \exp(i\xi\cdot A) = iA\cdot\omega_\mu(\xi) + iV\cdot\Theta_\mu(\xi), \tag{2.20}$$

where the coefficient functions $\omega_\mu{}^i(\xi)$, $\Theta_\mu{}^i(\xi)$ are so-called Cartan forms. Introducing the chiral covariant derivatives $D_\mu\xi_i \equiv \omega_\mu{}^i(\xi)$ and $D_\mu\Psi = [\partial_\mu + i\Theta_\mu(\xi)\cdot V]\,\Psi$, we finally arrive at (summation over repeated indices is understood)

$$\mathcal{L}(\Psi, \xi) = \overline{\Psi}(i\gamma_\mu D^\mu - M)\,\Psi - (\overline{\Psi}\gamma_\mu\gamma_5 V_i\Psi)\, D^\mu\xi_i + \mathcal{L}_0(\xi). \tag{2.21}$$

It is not difficult to prove that the meson Lagrangian $\mathcal{L}_0(\xi)$ may be expressed in terms of the covariant derivative of the meson fields as

$$\mathcal{L}_0(\xi) = \frac{F^2}{2} D_\mu\xi_i D^\mu\xi_i. \tag{2.22}$$

---

[4]) The σ particle has not been found experimentally.

In the following Section we will show how these considerations can be generalized in a straightforward way to construct $SU(N) \times SU(N)$ $(N = 3, 4, \ldots)$ invariant meson-baryon (quark) Lagrangians.

II.2. Covariant derivatives and $SU(3) \times SU(3)$ meson-baryon Lagrangian

In this Section we continue our introduction into the techniques of nonlinear Lagrangians based on nonlinear realizations of symmetry groups. Let us follow the work by Coleman, Wess and Zumino [7] and first study in more detail the transformation laws of the covariant derivative under a given group. This then enables one to construct invariant Lagrangians.

Consider the chiral group $G = SU(N) \times SU(N)$ with $H = SU(N)$ being its algebraic subgroup that leaves the vacuum invariant (in practical applications we shall always deal with the cases $N = 2, 3, 4$). The vector and axial vector generators $V_i = \lambda_i/2\, 1$, $A_i = \lambda_i/2\, \gamma_5 (i = 1, \ldots, N^2 - 1)$ of the Lie algebra satisfy the commutation relations and normalization conditions (cf. eq. (2.3))

$$[V_i, V_j] = i f_{ijk} V_k, \qquad [V_i, A_j] = i f_{ijk} A_k, \qquad [A_i, A_j] = i f_{ijk} V_k$$
$$\mathrm{Tr}\, V_i V_j = \mathrm{Tr}\, A_i A_j = 2\delta_{ij}, \qquad \mathrm{Tr}\, A_i V_j = 0, \tag{2.23}$$

where λ_i, f_{ijk} are the Gell-Mann matrices and antisymmetric structure constants of the group $SU(N)$, respectively, and the trace runs over internal and Lorentz indices. In some neighbourhood of the identity of G every group element $g \in G$ may then be decomposed uniquely into a product of the form

$$g = e^{i\alpha \cdot A} e^{i\beta \cdot V}, \qquad \left(\alpha \cdot A = \sum_{i=1}^{N^2-1} \alpha_i A_i \text{ etc.}\right). \tag{2.24}$$

Let further $\Phi_i (i = 1, \ldots, N^2 - 1)$ be the multiplet of pseudoscalar mesons associated with the adjoint representation of the subgroup $SU(N)$. The meson fields play here the role of Goldstone bosons corresponding to the spontaneously broken axial generators A_i of $SU(N) \times SU(N)$. They are subject to the following nonlinear transformation law which becomes a linear one when restricted to the subgroup H[5])

$$g e^{i\xi \cdot A} = e^{i\xi'(\xi, g) \cdot A} e^{iu(\xi, g) \cdot V}, \tag{2.25}$$

where $\xi_i = \Phi_i/F$, F being a parameter that turns out later on to be the meson decay constant.

Sometimes it will be also useful to consider the corresponding transformation law for the squared expression $\exp i2\xi \cdot A$. To this end, observe that the group G has an automorphism $R: g \to R(g)$ such that

$$V_i \to V_i, \qquad A_i \to -A_i.$$

Applying this automorphism to eq. (2.25), one readily gets

$$g e^{i2\xi \cdot A} R(g^{-1}) = e^{i2\xi' \cdot A}. \tag{2.26}$$

Thus, considering chiral transformations, $g = e^{i\alpha \cdot A}$ and $R(g) = g^{-1}$, the inverse of the transformation law (2.26) reproduces the transformation law (2.5) for the meson matrix $\Pi = -F e^{-i2\xi \cdot A}$ of the nonlinear σ-model.

---

5) Notice that eq. (2.25) defines a nonlinear transformation on the coset space $SU(N) \times SU(N)/SU(N)$.

Now, as to the transformation laws of fermion fields (baryons, quarks). Let Ψ be an arbitrary field belonging to some linear representation D of the algebraic subgroup H. Under a group transformation $g \in G$ it is assumed to transform as

$$\Psi' = D(e^{iu(\xi,g)\cdot V})\, \Psi = e^{iu(\xi,g)\cdot \overline{V}}\Psi, \tag{2.27}$$

where $\overline{V}_i$ are matrix generators of the linear representation D of H.
Thus far, about the field transformations. To construct chiral invariant Lagrangians, we need further the covariant derivatives of the fields, i.e. quantitaties which transform under G as representations of the algebraic subgroup H with ξ-dependent parameters. They are given in terms of Cartan forms by an expression analogous to eq. (2.20) or explicitly

$$\begin{aligned} D_\mu \xi_i &= -\frac{i}{2} \operatorname{Tr} \{A_i e^{-i\xi\cdot A}\, \partial_\mu e^{i\xi\cdot A}\}, \\ D_\mu \Psi &= \left(\partial_\mu + i\Theta_\mu(\xi)\cdot \overline{V}\right)\Psi, \\ \Theta_\mu{}^i(\xi) &= -\frac{i}{2} \operatorname{Tr} \{V_i e^{-i\xi\cdot A}\, \partial_\mu e^{i\xi\cdot A}\}. \end{aligned} \tag{2.28}$$

Their transformation under G is analogous to eq. (2.27), i.e.

$$\begin{aligned} (D_\mu\xi)' &= D^{(A)}(e^{iu(\xi,g)\cdot V})\, D_\mu\xi \\ (D_\mu\Psi)' &= D(e^{iu(\xi,g)\cdot V})\, D_\mu\Psi, \end{aligned} \tag{2.29}$$

where $D^{(A)}$ is the matrix representation defined by $A \cdot (D_\mu\xi)' = e^{iu\cdot V} A \cdot D_\mu \xi e^{-iu\cdot V}$ and $D(\ldots)$ is the linear representation defined in eq. (2.27). To prove the transformation laws (2.29) consider the transformed expression [18]

$$\begin{aligned} e^{-i\xi'\cdot A}\, \partial_\mu e^{i\xi'\cdot A} &= e^{iu(\xi,g)\cdot V} e^{-i\xi\cdot A} g^{-1}\, \partial_\mu (g e^{i\xi\cdot A} e^{-iu(\xi,g)\cdot V}) \\ &= e^{iu(\xi,g)\cdot V}(e^{-i\xi\cdot A}\, \partial_\mu e^{i\xi\cdot A})\, e^{-iu(\xi,g)\cdot V} + e^{iu(\xi,g)\cdot V}\, \partial_\mu e^{-iu(\xi,g)\cdot V}, \end{aligned}$$

where use has been made of the fact that g is independent of x. The Cartan forms $\omega_\mu{}^i(\xi) \equiv D_\mu\xi_i, \Theta_\mu{}^i(\xi)$ (cf. eqs. (2.20), (2.28)) transform then as

$$\begin{aligned} A \cdot (D_\mu\xi)' &= e^{iu(\xi,g)\cdot V} A \cdot D_\mu \xi e^{-iu(\xi,g)\cdot V} = A \cdot \{D^{(A)}(e^{iu(\xi,g)\cdot V})\, D_\mu\xi\}, \\ V \cdot \Theta_\mu(\xi)' &= e^{iu(\xi,g)\cdot V}\left(V \cdot \Theta_\mu(\xi) - i\,\partial_\mu\right) e^{-iu(\xi,g)\cdot V}. \end{aligned} \tag{2.30}$$

Rewriting the last equation in terms of the matrix generators $\overline{V}_i$ and taking into account the definition of $D_\mu\Psi$, eq. (2.28), yields, in fact, eqs. (2.29).
We are now ready to write down hadron Lagrangians by forming suitable invariant combinations of fields and covariant derivatives. Let us write

$$\mathcal{L}_{\text{Str}} = \mathcal{L}_{\text{inv}}(D_\mu\xi;\, \Psi, D_\mu\Psi) + \Delta\mathcal{L}(\xi, \Psi), \tag{2.31}$$

where $\mathcal{L}_{\text{inv}}$ denotes the $SU(N) \times SU(N)$ invariant part of the strong interacting Lagrangian, while $\Delta\mathcal{L}$ is a suitably chosen symmetry breaking term, the discussion of which will be postponed to Sect. III.2.1.
As a first illustration we consider the case of an $SU(3) \times SU(3)$ invariant meson-baryon Lagrangian ($N = 3$). Now $\Psi_i = B_i$ and $\xi_i (i = 1, \ldots, 8)$ denote the fields of the baryon

and meson octets and we have [19, 20] (summation over repeated indices is understood)

$$\mathscr{L}_{\text{inv}}(D_\mu\xi;\,B,\,D_\mu B) = \frac{F^2}{2}\,D_\mu\xi_i D^\mu\xi_i + \bar{B}(i\gamma_\mu D^\mu - M_0)\,B - g_A(\bar{B}_i\gamma_\mu\gamma_5 c_{ijk}B_j)\,D^\mu\xi_k. \quad (2.32)$$

Here $c_{ijk} = ad_{ijk} - i(1-a)\,f_{ijk}$, $a \approx 2/3$ is the mixing parameter of $F-D$ couplings, d_{ijk} are the symmetric structure constants of the group $SU(3)$, M_0 is an averaged baryon mass, and $g_A \approx 1.25$ is the axial vector renormalization constant. Expanding the covariant derivatives in the meson fields ξ, we also have

$$D_\mu\xi_i = \partial_\mu\xi_i - \frac{1}{6}\,f_{iln}f_{jmn}\xi_l\xi_m\,\partial_\mu\xi_j + O(\xi^5),$$

$$D_\mu B_i = \partial_\mu B_i - \frac{1}{2}\,f_{ijm}f_{klm}\,\partial_\mu\xi_k\xi_l B_j + O(\xi^4)\,B. \quad (2.28')$$

For practical reasons, it is useful to transform the Lagrangian (2.32) to get rid of the derivative term linear in the meson field ξ. To this end, we redefine the baryon fields by

$$B_i = \{\exp\left(-ig_A\gamma_5 c_{ijk}\xi_k\right)\}\,b_j. \quad (2.33)$$

This transformation leaves S-matrix elements unchanged. In the following we shall be interested only in those parts of the transformed Lagrangian which contain first and second powers in the meson fields. Inserting eq. (2.33) into (2.32) one receives the following three interaction Lagrangians.

$$\mathscr{L}_1 = i2gFc_{ijk}\bar{b}_i\gamma_5 b_j\xi_k, \quad (2.34\text{a})$$

$$\mathscr{L}_2 = 2\,\frac{g^2F^2}{M_0}\,c_{kjm}c_{lim}\bar{b}_j b_l\xi_i\xi_k, \quad (2.34\text{b})$$

$$\mathscr{L}_3 = \frac{i}{2}\,\bar{b}_i\gamma^\mu b_l\xi_j\,\partial_\mu\xi_k\Big\{f_{ilm}f_{kjm}(g_A{}^2 - 1) + g_A{}^2\left[\frac{2}{3}\,a^2(\delta_{ij}\delta_{kl} - \delta_{ik}\delta_{jl}) + 2a(a-1)\,f_{kjm}(f_{ilm} - id_{ilm})\right]\Big\}. \quad (2.34\text{c})$$

Note that $\mathscr{L}_1$ coincides with the standard $SU(3)$-expression for the pseudoscalar meson-baryon coupling, g being the strong interaction coupling constant ($g^2/4\pi \approx 14.7$). The latter one is related to g_A by the Goldberger-Treiman relation, $g \approx g_A M_0/F$ (or, in the case of symmetry breaking $g_{ij}^k = (g_A{}^k)_{ij}(M_i + M_j)/2F_k$).
For further generalizations we finally rewrite $\mathscr{L}_{\text{inv}}$ in terms of the familiar (3×3) baryon matrix B. Let

$$B = \sqrt{2}\,B_i V_i, \qquad \Theta_\mu = \Theta_\mu{}^i V_i,$$

then

$$\mathscr{L}_{\text{inv}} = \frac{F^2}{2}\,D_\mu\xi_i D^\mu\xi_i + \operatorname{Tr}\bar{B}(i\gamma_\mu D^\mu - M_0)\,B - g_A\operatorname{Tr}\left(\bar{B}\gamma_\mu\gamma_5(a\{V_i,\cdot\} + (1-a)\,[V_i,\cdot])\,B\right)D^\mu\xi_i, \quad (2.32')$$

where

$$D_\mu B = (\partial_\mu + i[\Theta_\mu,\cdot])\,B, \qquad ([V_i,\cdot]\,B \equiv [V_i, B]).$$

The transformation (2.33) takes now the form

$$B = \exp\left(-ig_A\gamma_5(a\{V_k,\cdot\} + (1-a)\,[V_k,\cdot])\,\xi_k\right)b.$$

II.3. Extension to $SU(4) \times SU(4)$ and chiral quark models

The recent discovery of new particles carrying another quantum number, charm, naturally suggests extending the chiral group $SU(3) \times SU(3)$ to $SU(4) \times SU(4)$. We shall now give a straight-forward extension of the Lagrangian (2.32) to this case [21]. Let ξ_i ($i = 1, 2, \ldots 15$) be the 15-plet of O^--mesons and $B_c^{[ab]}$ ($a, b, c, = 1 \ldots 4$) the tensor wave functions of mixed symmetry describing the 20-plet of $1/2^+$-baryons (for explicit expressions, cf. Appendix A)[6]. The $SU(4) \times SU(4)$ meson-baryon Lagrangian looks then formally like in eq. (2.32′) where now the sum over i runs over the 15-plet of mesons, and one has to use the following slightly generalized definitions of traces, commutators and anticommutators.

$$\mathrm{Tr}\,\bar{B}B = \frac{1}{2}\,\bar{B}^a_{[mn]} B_a^{[mn]},$$

$$[V_i, B]_a^{[mn]} \equiv (V_i)_a^{\ b}\, B_b^{[mn]} - \left(B_a^{[bn]} (V_i)_b^{\ m} + B_a^{[m,b]} (V_i)_b^{\ n}\right), \tag{2.35}$$

$$\{V_i, B\}_a^{[mn]} \equiv (V_i)_a^{\ b}\, B_b^{[mn]} + \left(B_a^{[bn]} (V_i)_b^{\ m} + B_a^{[mb]} (V_i)_b^{\ n}\right).$$

In writing down an $SU(4) \times SU(4)$ Lagrangian of the simple type (2.32′), it was assumed that there is no essential symmetry breaking in the axial renormalization constant g_A for charmed baryons. Assumptions of this kind must, of course, be justified or disproved by comparing with experiment.

Finally, for further applications, we introduce yet a second class of chiral models describing the interaction of mesons ξ_i and colored quarks q_{ia}. The quark indices refer here to color and flavor, respectively, $\left(i = 1, \ldots, 3,\ SU(3)_c;\ a = 1, \ldots, 4,\ SU(4)_f\right)$. The corresponding Lagrangian reads (cf. eqs. (2.21), (2.32))

$$\mathcal{L} = \frac{F^2}{2} D_\mu \xi_i D^\mu \xi_i + \bar{q}(i\gamma_\mu D^\mu - M_q^0)\, q - g_A'(\bar{q}\gamma_\mu\gamma_5 V_i q)\, D^\mu \xi_i;$$

$$q = \begin{pmatrix} u \\ d \\ s \\ c \end{pmatrix}, \tag{2.36}$$

where g_A' is the axial renormalization constant of quarks, and we have

$$D_\mu q = \left[\partial_\mu + \frac{i}{2} f_{klm} \xi_k\, \partial_\mu \xi_l V_m + O(\xi^4)\right] q.$$

In terms of new quark fields ψ, $q = \exp(-ig_A' \xi \cdot A)\, \psi$, we get

$$\mathcal{L} = \mathcal{L}_1 + \mathcal{L}_2 + \mathcal{L}_3 + \bar{\psi}\gamma_5 [O(\xi^3) + \gamma_\mu O(\xi^2\, \partial^\mu \xi)]\, \psi,$$

where (comp. eqs. 2.34a—c)

$$\mathcal{L}_1 = i2g'F(\bar{\psi}\gamma_5 V_k \psi)\, \xi_k, \tag{2.37a}$$

$$\mathcal{L}_2 = \frac{2g'^2 F^2}{M_q^0} (\bar{\psi} V_k V_l \psi)\, \xi_k \xi_l, \tag{2.37b}$$

---

[6]) The tensor wave functions $B_c^{[ab]}$ are given here in the quark representation. They transform as (comp. eq. (2.27)) $B_i^{[mn]\prime} = (U^+)_i^{\ m}\, (U^+)_j^{\ n}\, U_l^{\ k} B_k^{[ij]}$ where $U = \exp iu(\xi, g) \cdot V$.

$$\mathcal{L}_3 = \frac{(g_A'^2 - 1)}{2} f_{klm}(\bar{\psi}\gamma_\mu V_m \psi)\, \xi_k \partial^\mu \xi_l\,, \tag{2.37c}$$

$g' = g_A' M_q^0/F$ being the strong interaction meson-quark coupling constant. In the subsequent Section we shall sketch how the above techniques can be further extended to include gauge fields.

III. Construction of Unified Gauge Models

III.1. Gauge fields

So far we have studied only field transformations with respect to the global group $SU(N) \times SU(N)$. As is well-known, the construction of more general models containing besides hadron fields also gauge fields requires considering coordinate-dependent group transformations.

Let G_W be a subgroup of $G = SU(N) \times SU(N)$ with generators C_i $(i = 1, 2, \ldots, R)$ which are linear combinations of the generators V_i, A_i. Under local group transformations $h(x) \in G_W$ the hadron fields are supposed to transform as

$$\begin{aligned} &(\xi, \Psi) \to (\xi', \Psi') = h(x)\,(\xi, \Psi)\,, \\ &h(x) = e^{i\varepsilon(x)\cdot C} \in G_W\,, \qquad \left(\varepsilon \cdot C = \sum_{i=1}^{R} \varepsilon_i C_i\right) \end{aligned} \tag{3.1}$$

or explicitly (cf. eqs. (2.25), (2.27))

$$h(x)\, e^{i\xi\cdot A} = e^{i\xi'(\xi,h)\cdot A} e^{iu(\xi,h)\cdot V}$$

$$, \Psi' = D(e^{iu(\xi,h)\cdot V})\, \Psi\,.$$

As usual the construction of a Lagrangian invariant under the local group transformations (3.1) requires a set of gauge fields $W_\mu{}^i$ $(i = 1, 2, \ldots, R)$ associated to the generators of the local group G_W. Their transformation laws are

$$i\varkappa W_\mu' \cdot C = e^{i\varepsilon(x)\cdot C}(\partial_\mu + i\varkappa W_\mu \cdot C)\, e^{-i\varepsilon(x)\cdot C}\,, \tag{3.2}$$

where $\varkappa$ is a coupling constant which determines the strength of the universal coupling of the gauge fields to all other fields.

Let us further introduce gauge-covariant derivatives $\bar{D}_\mu \xi_i$, $\bar{D}_\mu \Psi$ of the hadron fields defined by

$$\begin{aligned} &e^{-i\xi\cdot A}(\partial_\mu + i\varkappa W_\mu \cdot C)\, e^{i\xi\cdot A} = iA \cdot \bar{D}_\mu \xi + iV \cdot \bar{\Theta}_\mu(\xi)\,, \\ &\bar{D}_\mu \Psi = \left(\partial_\mu + i\bar{\Theta}_\mu(\xi) \cdot \bar{V}\right) \Psi\,. \end{aligned} \tag{3.3}$$

It is then straightforward to verify that the gauge-covariant derivatives transform under the combined transformations (3.1), (3.2) in formally the same way as the covariant derivatives (2.29), (2.30),

$$\begin{aligned} &(\bar{D}_\mu \xi)' = D^{(A)}(e^{iu(\xi,h)\cdot V})\, \bar{D}_\mu \xi \\ &(\bar{D}_\mu \Psi)' = D(e^{iu(\xi,h)\cdot V})\, \bar{D}_\mu \Psi\,. \end{aligned} \tag{3.4}$$

We are now ready to write down nonlinear unified gauge Lagrangians. To this end, apply the "minimal substitution rule"

$$(D_\mu\xi, D_\mu\Psi) \to (\bar{D}_\mu\xi, \bar{D}_\mu\Psi) \tag{3.5}$$

to the strong interaction Lagrangian (2.31) and add the Lagrangian L_{YM} of the gauge (Yang-Mills) fields. This yields altogether

$$\mathcal{L} = \mathcal{L}_{\mathrm{inv}}(\bar{D}_\mu\xi; \Psi, \bar{D}_\mu\Psi) + \mathcal{L}_{\mathrm{YM}}(W_\mu) + \Delta\mathcal{L}(\xi, \Psi),$$

where

$$\mathcal{L}_{\mathrm{YM}}(W_\mu) = -\frac{1}{4}(W^k_{\mu\nu})^2, \qquad W^k_{\mu\nu} = \partial_\mu W_\nu{}^k - \partial_\nu W_\mu{}^k - \varkappa\omega^{klm} W_\mu{}^l W_\nu{}^m, \tag{3.6}$$

and ω^{klm} are the structure constants of the gauge group G_W. By construction, the sum $\mathcal{L}_{\mathrm{inv}}(D_\mu\xi; \Psi, \bar{D}_\mu\Psi) + \mathcal{L}_{\mathrm{YM}}$ is invariant under the group G_W.

III.2. Nonlinear realization of the Weinberg-Salam model

As an application of the above ideas we shall now demonstrate how one can introduce the electro-weak interactions into the chiral $SU(4) \times SU(4)$ meson-baryon (or meson-quark) Lagrangian of Sect. II.3. The gauge group we choose is the group $G_W = SU(2)_L \times U(1)$ of the standard Weinberg-Salam model [*22*].
To find the (nonlinear) transformation laws of the hadron fields (ξ, B) with respect to G_W it is convenient to first embed G_W into the global chiral group $SU(4) \times SU(4)$. The generators of G_W may then be represented by the following 4×4 matrices[7])

$$\begin{aligned} SU(2)_L\colon\ \hat{C}_i &= \frac{1+\gamma_5}{2}\frac{C_i}{2}, \qquad C_i = \begin{pmatrix} \tau_i & 0 \\ 0 & \tau_1\tau_i\tau_1^{-1} \end{pmatrix} \\ U(1)\colon\ \frac{\hat{Y}}{2} &= \frac{Y_W}{2} + \frac{1-\gamma_5}{2}\frac{C_3}{2}, \end{aligned} \tag{3.7}$$

where τ_i are Pauli matrices, and Y_W is a c-number to be fixed below. The matrix τ_1 in C_i is required by our choice of $SU(4)$ λ-matrices which act on the quark spinor.

$$q = \begin{pmatrix} u \\ d \\ s \\ c \end{pmatrix} \quad \text{but not} \quad \begin{pmatrix} u \\ d \\ c \\ s \end{pmatrix}. \tag{3.8}$$

Let us now determine the c-number Y_W. To this end, we consider the operator of the electromagnetic charge Q defined as usually by the sum of the third component of the weak isospin $\hat{C}_3$ and the weak hypercharge $\hat{Y}/2$

$$Q = \hat{C}_3 + \frac{\hat{Y}}{2} = \frac{1}{2}(C_3 + Y_W). \tag{3.9}$$

Choosing $Y_W = 1/3$ yields in fact the usual quark charges, diag $Q = (2/3, -1/3, -1/3, 2/3)$. (On the other hand, for the lepton or Higgs boson multiplets one gets $Y_W = -1$ or $+1$, respectively (see Appendix B)).

[7]) Expressed in terms of these generators the usual lepton sector of the Weinberg-Salam model can be written in a compact fourdimensional form (cf. Appendix B and eq. (3.41)).

We then postulate the following transformation laws[8]) ($Y_W = 1/3$)

$$h(x)\, e^{i\xi\cdot A} = e^{i\eta(x)(Y_W/2)}(e^{i\xi'\cdot A} e^{iu(\xi,h)\cdot V})$$

$$B_l^{[mn]'} = e^{i3\eta(x)(Y_W/2)}(U^+)_i^{\ m}\,(U^+)_j^{\ n}\,U_l^{\ k} B_k^{[ij]},$$

or

$$q_l' = e^{i\eta(x)(Y_W/2)} U_l^{\ k} q_k\,, \tag{3.10}$$

where

$$h(x) = e^{i\eta(x)(\hat{Y}/2)} e^{i\varepsilon(x)\cdot\hat{C}} \in G_W$$

$$U = e^{iu(\xi,h)\cdot V}.$$

Furthermore, one must now introduce two sorts of gauge bosons $W_\mu^{\ i}$ and X_μ, corresponding to the two sets of generators $\hat{C}_i$ and $\hat{Y}/2$ of the subgroups $SU(2)_L$ and $U(1)$ of G_W. Their respective transformation laws are (cf. eq. (3.2))

$$i\varkappa W_\mu' \cdot \hat{C} = e^{i\varepsilon(x)\cdot\hat{C}}(\partial_\mu + i\varkappa W_\mu \cdot \hat{C})\, e^{-i\varepsilon(x)\cdot\hat{C}},$$
$$i\varrho X_\mu' = i\varrho X_\mu - i\partial_\mu \eta(x)\,. \tag{3.11}$$

Notice that the non-Abelian and Abelian gauge bosons $W_\mu^{\ i}$ and X_μ couple with different coupling constants $\varkappa$ or ϱ, respectively.

The gauge-covariant derivatives $\bar{D}_\mu\xi$, $\bar{D}_\mu B$ or $\bar{D}_\mu q$ associated to the "electro-weak" group $SU(2)_L \times U(1)$ are [21, 23, 24] (comp. eq. (3.3))

$$\begin{pmatrix} \bar{D}_\mu \xi_i \\ \bar{\Theta}_\mu^{\ i}(\xi) \end{pmatrix} = -\frac{i}{2}\,\mathrm{Tr}\left\{\begin{pmatrix} A_i \\ V_i \end{pmatrix} e^{-i\xi\cdot A}(\partial_\mu + i\varkappa W_\mu \cdot \hat{C} + i\varrho X_\mu \hat{Y}/2)\, e^{i\xi\cdot A}\right\} \tag{3.12}$$

$$\bar{D}_\mu B = \left(\partial_\mu + i3\,\frac{Y_W}{2}\,\varrho X_\mu + i[\bar{\Theta}_\mu, \cdot]\right) B, \tag{3.13}$$

or

$$\bar{D}_\mu q = \left(\partial_\mu + i\,\frac{Y_W}{2}\,\varrho X_\mu + i\bar{\Theta}_\mu \cdot V\right) q, \tag{3.14}$$

respectively. Applying further the minimal substitution rule (3.5) to the strong interaction Lagrangians $\mathscr{L}_{\mathrm{inv}}(D_\mu\xi;\, \Psi, D_\mu\Psi)$ quoted in Sect. II.3, we arrive at $(\Psi \equiv (B, q))$

$$\mathscr{L}_{\mathrm{inv}}(\bar{D}_\mu\xi;\, \Psi, \bar{D}_\mu\Psi) = \mathscr{L}_{\mathrm{inv}}(D_\mu\xi;\, \Psi, D_\mu\Psi)$$
$$- \frac{\varkappa}{2\sqrt{2}}\,(W^{\mu +} J^W_{\mu,h} + \text{h.c.}) - \frac{\varkappa}{2\cos\Theta_W}\, Z^\mu J^Z_{\mu,h}$$
$$- eA^\mu J^A_{\mu,h} + \mathscr{L}_{\mathrm{bil}}(A_\mu, W_\mu^{\ \pm}, Z_\mu)\,. \tag{3.15}$$

The expressions for the currents J_μ and the bilinear term $\mathscr{L}_{\mathrm{bil}}$ will be given below and in Appendix C, respectively. Obviously, the gauge-invariant Lagrangian reproduces the strong interaction Lagrangian we startet with. In addition, there arise terms describing the interaction of the hadrons with the fields $W_\mu^{\ \pm}$, Z_μ and A_μ of the charged and neutral vector bosons and photon, respectively.

---

[8]) Strictly speaking, the group G_W must be embedded into the group $U(4) \times U(4)$ since the operator of the weak hypercharge $\hat{Y}/2$ contains the unit matrix. The latter leads to additional phase factors in the transformation laws of the fields (see (3.10)) and to additional terms in the covariant derivatives eqs. (3.13), (3.14).

The physical gauge bosons are defined by

$$W_\mu^\pm = \frac{1}{\sqrt{2}} (W_\mu^1 \mp iW_\mu^2), \qquad Z_\mu = \cos\Theta_W W_\mu^3 - \sin\Theta_W X_\mu,$$
$$A_\mu = \sin\Theta_\mu W_\mu^3 + \cos\Theta_W X_\mu, \tag{3.16}$$

Θ_W being the Weinberg angle, $\Theta_W = \arctan \varrho/\varkappa$, and e the electromagnetic charge, $e = \varkappa \sin\Theta_W$ [22]. The weak charged and neutral hadronic currents $J^W_{\mu,h}$, $J^Z_{\mu,h}$ and the electromagnetic current $J^A_{\mu,h}$ are $\left(V_0 = 1/2\sqrt{2}\,1\right)$ [21, 23]

$$\begin{aligned} J_{\mu,h} = {} & -F^2\bar{D}_\mu\xi_i\langle PA_i(\xi)\rangle + \{\langle\bar{B}\gamma_\mu[V_i, B]\rangle\langle PV_i(\xi)\rangle \\ & + \langle\bar{B}\gamma_\mu V_0 B\rangle\langle 3PV_0\rangle + g_A\langle\bar{B}\gamma_\mu\gamma_5(a\{V_i, B\} + (1-a)[V_i, B])\rangle\langle PA_i(\xi)\rangle\}, \end{aligned} \tag{3.17}$$

where P is a projection operator defined by

$$P = \{2\hat{C}_+, 2\hat{C}_3 - 2\sin^2\Theta_W Q, Q\}$$

for $\{J_\mu{}^W, J_\mu{}^Z, J_\mu{}^A\}$, respectively, and $\hat{C}_\pm = \hat{C}_1 \pm i\hat{C}_2$. Henceforth we use the short-hand notations $\langle\bar{A}B\rangle = 1/2\,\bar{A}^a_{[mn]}B_a{}^{[mn]}$ for baryon-like tensors, $\langle MN\rangle = 1/2\,M_j{}^iN_i{}^j$ (including the trace over possible Lorentz indices) for meson matrices and

$$\begin{pmatrix} A_i(\xi) \\ V_i(\xi) \end{pmatrix} = e^{i\xi\cdot A}\begin{pmatrix} A_i \\ V_i \end{pmatrix} e^{-i\xi\cdot A}. \tag{3.18}$$

Analogously, the currents of the meson-quark model are

$$J_\mu = -F^2\bar{D}_\mu\xi_i\langle PA_i(\xi)\rangle + \left\{\sum_{i=0}^{15}(\bar{q}\gamma_\mu V_i q)\langle PV_i(\xi)\rangle + g_A'\sum_{i=1}^{15}(\bar{q}\gamma_\mu\gamma_5 V_i q)\langle PA_i(\xi)\rangle\right\}. \tag{3.19}$$

Notice that, if $\Delta\mathcal{L} = 0$, the currents (3.17), (3.19) are just the conserved Noether currents of the group $SU(2)_L \times U(1)$ (for details see Sect. III. 2.2). In Appendix D some useful formulas for the meson currents can be found.

III. 2.1. $SU(4) \times SU(4)$ breaking, Cabibbo angle and mass formulas

As we have seen, the chiral invariant strong interaction Lagrangian describes massless (Goldstone) mesons and equal-mass baryons (quarks). In order to get a physical hadron mass-spectrum, one must further add a symmetry breaking term $\Delta\mathcal{L}(\xi, \Psi)$.
Let us consider a $SU(4) \times SU(4)$ breaking mechanism that fulfils the following requirements [25]. It contains a minimal number of arbitrary parameters, reproduces a qualitatively correct picture of the masses of the 15-plet of 0^--mesons and 20-plet of $^1/_2{}^+$-baryons (with an uncertainty of, say, 10%) and yields the Cabibbo structure of the charged weak current including its PCAC condition. Moreover, there is a relationship between the strength of the $SU(2) \times SU(2)$ breaking (the finiteness of the pion mass) and the nonconservation of strangeness in weak interaction which is realized by the Cabibbo angle.
We remind the reader that a symmetry breaking scheme with an analogous set of requirements has been proposed for the chiral group $SU(3) \times SU(3)$ by Oakes many years ago [26]. There exist also attempts in the literature to modify and improve it [27]. However, as has been shown in ref. [2, 28], its original version suffices for practical purposes and provides a satisfactory description of low-energy interactions of hadrons.

The idea behind this approach is the following. The group $SU(4)\times SU(4)$ (or $SU(3)\times SU(3)$) is first broken down to $SU(2)\times SU(2)$. Then, the breaking of $SU(2)\times SU(2)$ and thus a finite pion mass is achieved by a mixing of the strange and nonstrange hadrons where the mixing angle turns out to be the Cabibbo angle.

Meson masses

Let us now illustrate how the method works by considering the symmetry breaking mechanism generating the meson masses. The respective symmetry breaking term $\Delta\mathcal{L}_m$ is usually assumed to transform predominantly as a sum of terms of the representation $(4,4^*)+(4^*,4)$. We have

$$\Delta\mathcal{L}_m=\frac{\mu^2}{\sqrt{8}}F^2\langle(V_0+aV_8+bV_{15})\,e^{i2\xi\cdot\Lambda}\rangle. \tag{3.20}$$

To preserve $SU(2)\times SU(2)$ symmetry, one must relate the parameters μ^2, a, b to the masses of the neutral K and D mesons as follows

$$\mu^2=m_{K^0}^2+m_{D^0}^2,\qquad a=-\sqrt{\frac{8}{3}}\,m_{K^0}^2/\mu^2$$

$$b=-(3m_{D^0}^2-m_{K^0}^2)/\sqrt{3}\,\mu^2.$$

In the next step, $SU(2)\times SU(2)$ will be broken, too, by rotating the "bare" fields ξ_i around the seventh axis in $SU(4)$ space. This yields, finally,

$$\Delta\mathcal{L}_m=\frac{F^2}{2}\left\langle U_c^+\begin{pmatrix}0&0&0&0\\0&0&0&0\\0&0&m_{K^0}^2&0\\0&0&0&m_{D^0}^2\end{pmatrix}U_c e^{i2\bar{\xi}\cdot\Lambda}\right\rangle_{\Delta S=0}, \tag{3.21}$$

or

$$\Delta\mathcal{L}_m=-\frac{F^2}{2}\{m_{K^0}^2\sin^2\Theta(\bar{\xi}\cdot\lambda)_{22}^2+m_{K^0}^2\cos^2\Theta(\bar{\xi}\cdot\lambda)_{33}^2+m_{D^0}^2(\bar{\xi}\cdot\lambda)_{44}^2\}+O(\xi^4),$$

where the new "physical" fields $\bar{\xi}$ are defined by

$$e^{i2\bar{\xi}\cdot\Lambda}=U_c^+e^{i2\xi\cdot\Lambda}U_c,\qquad U_c=e^{i2\Theta V_7} \tag{3.22}$$

and we have retained in eq. (3.21) only the strangeness-conserving parts. The mass formulas for the 15-plet of pseudoscalar mesons then read

$$m_{\pi^+}^2=m_{K^0}^2\sin^2\Theta,\qquad m_{K^+}^2=m_{K^0}^2\cos^2\Theta$$

$$m_{D^+}^2=m_{D^0}^2+m_{\pi^+}^2,\qquad m_{F^+}^2=m_{D^0}^2+m_{K^+}^2 \tag{3.23}$$

$$m_{\eta_8}^2=m_{K^0}^2\left(\frac{1}{3}+\cos^2\Theta\right),\qquad m_{\eta_{15}}^2=\frac{1}{6}(m_{K^0}^2+9m_{D^0}^2).$$

Choosing now $\Theta\approx 0.26$, $m_{K^0}=498$ MeV, $m_{D^0}=1863$ MeV yields the mass values $m_{\pi^+}=128$ MeV, $m_{K^+}=481$ MeV, $m_{D^+}=1867$ MeV and $m_{F^+}=1924$ MeV that are in rough agreement with experiment.

Eq. (3.21) contains a further term

$$-\frac{m_{\pi^+}^2}{\sqrt{3}}\pi^0\left(\eta_8+\frac{\eta_{15}}{\sqrt{2}}\right)-\frac{(3m_{K^+}^2-m_{K^0}^2)}{3\sqrt{2}}\eta_8\eta_{15} \tag{3.24}$$

leading to $\pi^0-\eta_8-\eta_{15}$ mixing. Instead of diagonalizing the full $\pi^0-\eta_8-\eta_{15}$ mass matrix it suffices for our purposes to diagonalize the $\pi^0-\eta_8$ mixing only (the $\pi^0-\eta_{15}$ and $\eta_8-\eta_{15}$ mixing terms give very small corrections due to large masses m_{η_8} and $m_{\eta_{15}}$). By introducing new fields

$$\pi^0=\pi^{0\prime}\cos\varphi+\eta_8{}'\sin\varphi$$

$$\eta_8=-\pi^{0\prime}\sin\varphi+\eta_8{}'\cos\varphi \tag{3.25}$$

one obtains $m_{\pi^{0\prime}}=127$ MeV, $m_{\eta_8'}=559$ MeV and $\varphi\approx-0.03$. The mass value $m_{\eta_{15}}=2291$ MeV is rather small when confronted with the mass of a recently reported particle $\psi_P(m_{\psi_P}=2.98\pm0.02$ GeV) which is considered to be a possible condidate for the η_c meson [*29*]. The correct treatment of the masses of the η,η' and η_c mesons requires at least also the inclusion of the $SU(4)$-singlet η_0. A detailed study of the $\eta-\eta'-\eta_c$ puzzle can be found in ref. [*30*].
Notice that the nonelectromagnetic mass splittings obtained above are in general too large (except the $\pi^\pm\pi^0$ mass difference) although their signs are correct. On the other hand, as is wellknown [*31*], pure electromagnetic corrections are unable to reproduce even the correct sign of the mass splittings if one does not admit sophisticated tadpole terms which are related to the exchange of hypothetical scalar mesons. If we combine the above nonelectromagnetic mass splittings with the electromagnetic mass corrections calculated in ref. [*31*], the $\pi^\pm\pi^0$ mass difference turns out to be almost completely of electromagnetic nature. Also the other too large mass differences are reduced, although not enough. The nonelectromagnetic mass splittings can, of course, be further improved by choosing modified symmetry breaking schemes. Anyway, the price one has then to pay is the appearance of new free parameters and loss of simplicity.

Baryon masses

To get baryon mass formulas let us choose the following symmetry breaking term close in spirit to eq. (3.21) [*25*]

$$\Delta\mathcal{L}_b=\frac{1}{2}\bar{\tilde{B}}^i_{[mn]}\begin{pmatrix}0&&&\\&a\sin^2\Theta&&\\&&a\cos^2\Theta&\\&&&b\end{pmatrix}_i^j\tilde{B}_j{}^{[mn]}+\bar{\tilde{B}}^m_{[in]}\begin{pmatrix}0&&&\\&c\sin^2\Theta&&\\&&c\cos^2\Theta&\\&&&d\end{pmatrix}_j^i\tilde{B}_m{}^{[jn]}. \tag{3.26}$$

Here $\tilde{B}_l{}^{[mn]}=(U_c{}^+)_i{}^m\,(U_c{}^+)_j{}^n\,(U_c)_l{}^k\,B_k{}^{[ij]}$ are "physical" baryon fields, U_c being the Cabibbo rotation defined in eq. (3.22). In writing (3.26) the possibility of two different baryon couplings (D—F) has been accounted for.
Eq. (3.26) yields the following baryon mass formulas (the notations for charmed baryons are as in ref. [*32*])

$$M_P=(M_0-d)-c\cos^2\Theta,\qquad M_{\Sigma^+}=(M_0-d)-c\sin^2\Theta,$$

$$M_N=M_P-a\sin^2\Theta,\qquad M_{\Sigma^0}=(M_0-d)-\frac{a+c}{2}\sin^2\Theta, \tag{3.27}$$

$$M_\Lambda = (M_0 - d) - \frac{a+c}{6}(1 + 3\cos^2\Theta), \qquad M_{\Sigma^-} = (M_0 - d) - a\sin^2\Theta,$$

$$M_{\Xi^0} = (M_0 - d) - c\sin^2\Theta - a\cos^2\Theta,$$

$$M_{\Xi^-} = (M_0 - d) - a\cos^2\Theta,$$

$$M_{A_1^+} = M_0 - \frac{2}{3}(b + d) - \frac{a+c}{6}\sin^2\Theta - c\cos^2\Theta,$$

$$M_{A_2^+} = M_0 - \frac{2}{3}(b + d) - \frac{a+c}{6}\cos^2\Theta - c\sin^2\Theta,$$

$$M_{A_2^0} = M_0 - \frac{2}{3}(b + d) - \frac{a+c}{6},$$

$$M_{B_3^{++}} = M_0 - c, \qquad M_{B_3^+} = M_0 - c - \frac{a-c}{2}\sin^2\Theta,$$

$$M_{B_3^0} = M_0 - c\cos^2\Theta - a\sin^2\Theta,$$

$$M_{B_2^+} = M_0 - c - \frac{a-c}{2}\cos^2\Theta, \qquad M_{B_2^0} = M_0 - \frac{a+c}{2}, \tag{3.27}$$

$$M_{B_1^0} = M_0 - c\sin^2\Theta - a\cos^2\Theta,$$

$$M_{C_2^{++}} = M_0 - b - c, \qquad M_{C_2^+} = M_0 - c\cos^2\Theta - b,$$

$$M_{C_1^+} = M_0 - c\sin^2\Theta - b.$$

Since $\Delta\mathcal{L}_b$ contains the mixing term

$$\frac{c+a}{2\sqrt{3}}\cos^2\Theta\{\bar{B}_2^+A_2^+ + \bar{B}_2^0A_2^0 + \text{h.c.}\}$$
$$- \frac{c+a}{2\sqrt{3}}\sin^2\Theta\{\bar{B}_2^0A_2^0 + \bar{B}_3^+A_1^+ + \bar{\Sigma}^0\Lambda + \text{h.c.}\}, \tag{3.28}$$

some of the states appearing in eq. (3.27) have yet to be diagonalized.
Using the three relations following from eq. (3.27)

$$M_P + M_{\Sigma^+} = 2M - c, \quad M_{\Xi^-} + M_{\Sigma^-} = 2M - a, \quad M_{\Xi^0} + M_N = 2M - a - c,$$

where $M = M_0 - d$, as well as the mass formulas for the A_1^+ and B_3^{++} baryons with the recently obtained mass values $M_{A_1^+} = 2260$ MeV, $M_{B_3^{++}} = 2426$ MeV [33] we may fix the 5 parameters M_0, a, b, c and d as follows

$$a = -127\ \text{MeV}, \qquad b = -1221\ \text{MeV}$$
$$c = 264\ \text{MeV}, \qquad d = 1494\ \text{MeV}$$
$$M_0 = 2690\ \text{MeV}.$$

Note that the parameter M_0 has no longer the meaning of an averaged baryon mass because the chosen $\Delta\mathcal{L}_b$ is asymmetric.

Table 1 contains the masses of the 20-plet of $^1/_2{}^+$-baryons obtained from eqs. (3.27), (3.28) after having performed the necessary diagonalizations. By analogous considerations one can also get quark mass relations.

Table 1

Mass values for the 20-plet of $^1/_2{}^+$-baryons

notations		$SU(3)$	(I, I_3)	S	C	mass (MeV)
P		8	$(^1/_2, {}^1/_2)$	0	0	949
N		8	$(^1/_2, -{}^1/_2)$	0	0	958
Λ		8	(0, 0)	−1	0	1109
Σ^+		8	(1, 1)	−1	0	1179
Σ^0		8	(1, 0)	−1	0	1191
Σ^-		8	(1, −1)	−1	0	1204
Ξ^0		8	$(^1/_2, {}^1/_2)$	−2	0	1297
Ξ^-		8	$(^1/_2, -{}^1/_2)$	−2	0	1315
$A_1{}^{+}$[32]	$C_0{}^{+}$[85]	$\bar{3}$	(0, 0)	0	1	2260
$A_2{}^+$	A^+	$\bar{3}$	$(^1/_2, {}^1/_2)$	−1	1	2460
$A_2{}^0$	A^0	$\bar{3}$	$(^1/_2 - {}^1/_2)$	−1	1	2476
$B_3{}^{++}$	$C_1{}^{++}$	6	(1, 1)	0	1	2426
$B_3{}^+$	$C_1{}^+$	6	(1, 0)	0	1	2439
$B_3{}^0$	$C_1{}^0$	6	(1, −1)	0	1	2452
$B_2{}^+$	S^+	6	$(^1/_2, {}^1/_2)$	−1	1	2618
$B_2{}^0$	S^0	6	$(^1/_2, -{}^1/_2)$	−1	1	2629
$B_1{}^0$	T^0	6	(0, 0)	−2	1	2791
$C_2{}^{++}$	$X_u{}^{++}$	3	$(^1/_2, {}^1/_2)$	0	2	3647
$C_2{}^+$	$X_d{}^+$	3	$(^1/_2, -{}^1/_2)$	0	2	3664
$C_1{}^+$	$X_s{}^+$	3	(0, 0)	−1	2	3894

Let us now rewrite the unified Lagrangian (3.15) in terms of physical fields $(\tilde{\xi}, \tilde{\Psi})$. Clearly, one has

$$\mathscr{L}_{\text{inv}}(D_\mu\xi; \Psi, D_\mu\Psi) = \mathscr{L}_{\text{inv}}(D_\mu\tilde{\xi}; \tilde{\Psi}, D_\mu\tilde{\Psi}).$$

Now, as for the currents. They retain their general form given by eqs. (3.17), (3.19) with the difference, however, that first, we must use a "rotated" projection operator $\tilde{P}$,

$$\tilde{P} = \{2\hat{C}_+(\Theta), 2\hat{C}_3 - 2\sin^2\Theta_W Q, Q\}, \qquad \hat{C}_+(\Theta) = U_c{}^+\hat{C}_+U_c$$

and, second, $\bar{D}_\mu\tilde{\xi}_i$ is defined by an equation analogous to (3.12) where ξ_i and $\hat{C}_i$ are now replaced by $\tilde{\xi}_i$, $\hat{C}_i(\Theta)$.

As promised above, let us next show that the angle Θ appearing in the symmetry breaking term of the Lagrangian coincides indeed with the Cabibbo angle of the charged weak hadronic current. To this end, decompose the charged projection operator $\tilde{P}^{\text{ch}} = 2\hat{C}_+(\Theta)$ as follows

$$\begin{aligned}\tilde{P}^{\text{ch}} &= \cos\Theta[(V^{1+i2} + A^{1+i2}) + (V^{13-i14} + A^{13-i14})] \\ &\quad + \sin\Theta[(V^{4+i5} + A^{4+i5}) - (V^{11-i12} + A^{11-i12})], \\ V^{k\pm il} &= V_k \pm iV_l \text{ etc.}\end{aligned} \tag{3.29}$$

Inserting eq. (3.29) into the expression for the charged weak current $J^W_{\mu,h}$ one gets in fact the following generalized Cabibbo formula of the GIM-scheme [34]

$$J^W_{\mu,h} = \cos \Theta(J^{1+i2}_{\mu,h} + J^{13-i14}_{\mu,h}) + \sin \Theta(J^{4+i5}_{\mu,h} - J^{11-i12}_{\mu,h}). \tag{3.30}$$

This way, we arrived at the remarkable result that the $SU(4)$ rotation originally introduced in the mass terms in order to break chiral $SU(2) \times SU(2)$ (and thus to generate a finite pion mass) also relates the different charged weak hadronic currents. Indeed, the correct form of the ordinary $SU(3)$ weak hadronic current will be reproduced by identifying the mixing angle Θ with the Cabibbo angle. Notice that the currents have the usual $V + A$ forms. $J^{1+i2}_{\mu,h}$, $J^{13-i14}_{\mu,h}$, $J^{4+i5}_{\mu,h}$ and $J^{11-i12}_{\mu,h}$ describe weak transitions with the following changes of strangeness S and charm C, respectively: $\Delta S = \Delta C = 0$, $\Delta S = \Delta C = 1$, $\Delta S = 1 \neq \Delta C = 0$ and $\Delta S = 0 \neq \Delta C = 1$.
In conclusion, we mention that the neutralcurrent $J^Z_{\mu,h}$ admits the well-known decomposition

$$J^Z_{\mu,h} = J^{(2\hat{C}_3)}_{\mu,h} - 2 \sin^2 \Theta_W J^A_{\mu,h}, \tag{3.31}$$

where $J^{(2\hat{C}_3)}_{\mu,h}$ is the current belonging to the third component of the weak isospin.

III.2.2. PCAC relations

In this subsection we shall outline how the currents (3.17) and their PCAC relations can be derived. Following Gell-Mann and Levy [4], one considers for a given group G, infinitesimal field transformations with coordinate-dependent parameters $\varepsilon_i(x)$

$$\Phi_j(x) \to \Phi_j(x) + \varepsilon_k(x)\, F_j{}^k(\Phi), \tag{3.32}$$

where the function $F_j{}^k(\Phi)$ may, in general, depend nonlinearly on the fields Φ_i. Under the field variations (3.32) the Lagrangian undergoes the change

$$\delta \mathcal{L} = \frac{\delta \mathcal{L}}{\partial \varepsilon_k} \varepsilon_k + \frac{\delta \mathcal{L}}{\delta(\partial_\mu \varepsilon_k)} \partial_\mu \varepsilon_k, \tag{3.33}$$

with

$$\begin{aligned} \frac{\delta \mathcal{L}}{\delta \varepsilon_k} &= \frac{\delta \mathcal{L}}{\delta \Phi_j} F_k{}^j + \frac{\delta \mathcal{L}}{\delta(\partial_\mu \Phi_j)} \partial_\mu F_k{}^j \\ \frac{\delta \mathcal{L}}{\delta(\partial_\mu \varepsilon_k)} &= \frac{\delta \mathcal{L}}{\delta(\partial_\mu \Phi_j)} F_k{}^j. \end{aligned} \tag{3.34}$$

The currents are then defined by (see also (2.6))

$$J_{k\mu} \equiv -\frac{\delta \mathcal{L}}{\delta(\partial^\mu \varepsilon_k)} = -\frac{\delta \mathcal{L}}{\delta(\partial^\mu \Phi_j)} F_k{}^j(\Phi). \tag{3.35}$$

Finally, taking into account the Euler-Lagrange equations

$$\frac{\delta \mathcal{L}}{\delta \Phi_i} = \partial_\mu \frac{\delta \mathcal{L}}{\delta(\partial_\mu \Phi_i)}, \tag{3.36}$$

and eqs. (3.34), (3.35), one gets the divergence relations

$$\partial^\mu J_{k\mu} = -\frac{\delta \mathcal{L}}{\delta \varepsilon_k}. \tag{3.37}$$

Thus, if the Lagrangian is invariant under the transformations (3.32) (now with constant ε_k) the currents are conserved, $\partial^\mu J_{k\mu} = 0$.
Before applying this method to our situation, it is convenient to include, for completeness, also the lepton sector of the standard Weinberg-Salam model into the total Lagrangian (cf. Appendix B). We have

$$\mathcal{L}_{\text{tot}} = \mathcal{L}_{\text{inv}}(\bar{D}_\mu \tilde{\xi}; \tilde{B}, \bar{D}_\mu \tilde{B}) + \mathcal{L}_{\text{WS}} + \Delta\mathcal{L}(\tilde{\xi}, \tilde{B}), \tag{3.38}$$

where the first term is the invariant Lagrangian (3.15) expressed in terms of "physical" fields $\tilde{\xi}$, $\tilde{B}$ (the discussion for quark fields $\tilde{q}$ is analogous), $\mathcal{L}_{\text{WS}}$ is the Lagrangian of the leptonsector of the Weinberg-Salam model and $\Delta\mathcal{L}$ denotes a symmetry-breaking term. Let us consider the following variations of the hadron fields $(\tilde{\xi}, \tilde{B})$ and of the lepton fields l (cf. eqs. (3.10))

$$\delta e^{i\tilde{\xi}\cdot A} \sim U_c^+ \left(\Pi - i\eta \frac{Y_W}{2}\right) U_c e^{i\tilde{\xi}\cdot A}$$

$$\delta\tilde{B} \sim i3\eta \frac{Y_W}{2} \tilde{B} \tag{3.39}$$

$$\delta l = \Pi l, \quad l = \begin{pmatrix} \nu_e \\ e \\ \mu \\ \nu_\mu \end{pmatrix}$$

where

$$\begin{aligned} \Pi &= i\left(\sum_{i=1}^{3} \varepsilon_i(x)\, \hat{C}_i + \eta(x) \frac{\hat{Y}}{2}\right) \\ &= i\,[(\varepsilon_+' 2\hat{C}_+ + \varepsilon_-' \, 2\hat{C}_-) + \varepsilon_A' Q + \varepsilon_Z'(2\hat{C}_3 - 2\sin^2\Theta_W Q)]. \end{aligned}$$

The sign $\sim$ in the transformation law of the hadron fields means "up to the terms of the order $O(u \cdot V)$". Such terms will be dropped here because they either cancel each other in the Lagrangian or contribute only to the divergence of the vector current but not to the PCAC relation. To reproduce the standard normalization for the lepton currents we have introduced slightly modified ε parameters. Their defining relations are

$$\begin{aligned} \varepsilon_3 &\equiv \varepsilon_A' + 2\cos^2\Theta_W \varepsilon_Z', \\ \eta &\equiv \varepsilon_A' - 2\sin^2\Theta_W \varepsilon_Z', \\ \frac{1}{\sqrt{2}}(\varepsilon_1 \mp i\varepsilon_2) &\equiv 2\sqrt{2}\,\varepsilon_\pm'. \end{aligned} \tag{3.40}$$

With these conventions, the lepton currents are

$$\begin{aligned} J^W_{\mu,l} &= -\frac{\delta(\mathcal{L}_{\text{WS}})_l}{\delta(\partial^\mu \varepsilon_+')} = \bar{l}\gamma_\mu(2\hat{C}_+)\, l, \\ J^Z_{\mu,l} &= -\frac{\delta(\mathcal{L}_{\text{WS}})_l}{\delta(\partial^\mu \varepsilon_Z')} = \bar{l}\gamma_\mu(2\hat{C}_3 - 2\sin^2\Theta_W Q)\, l, \\ J^A_{\mu,l} &= \bar{l}\gamma_\mu Q l, \end{aligned} \tag{3.41}$$

where $(\mathcal{L}_{\text{WS}})_l$ is the lepton part of the Lagrangian $\mathcal{L}_{\text{WS}}$.

The total currents get contributions from the hadron fields (contained in $\mathcal{L}_{\text{inv}}$) and from the lepton and gauge fields (contained in $\mathcal{L}_{\text{WS}}$). Note that in order to calculate them one must of course take into account also the variations $\delta W_\mu{}^i$ and δX_μ of the gauge fields[9]). It is not difficult to show that applying the field variations (3.39) to the Lagrangian $\mathcal{L}_{\text{inv}}$ in eq. (3.38) and using the definition (3.35) yields indeed the hadronic currents (3.17). Let us next come to the divergence relations for the total currents $J_\mu{}^{\text{tot}}$. Using eqs. (3.37), (3.38) and the fact that the first two terms in eq. (3.38) are invariant, we have

$$\partial^\mu J_\mu{}^{\text{tot}} = -\frac{\delta \Delta \mathcal{L}}{\delta \varepsilon'}. \tag{3.42}$$

To calculate $\delta \Delta \mathcal{L}$ it is simplest to use the transformation law (2.26) for the square of the meson matrix, i.e. $\exp(i2\xi \cdot A)$. Its infinitesimal version, written in terms of physical fields, reads

$$\begin{aligned} \delta_+ e^{i2\xi\cdot A} &= i\varepsilon_+'\left([C_+(\Theta), e^{i2\xi\cdot A}] + \gamma_5\{C_+(\Theta), e^{i2\xi\cdot A}\}\right), \\ \delta_Z e^{i2\xi\cdot A} &= i\varepsilon_Z'\left((1 - 2\sin^2\Theta_W)\left[\frac{C_3}{2}, e^{i2\xi\cdot A}\right] + \gamma_5\left\{\frac{C_3}{2}, e^{i2\xi\cdot A}\right\}\right), \\ \delta_A e^{i2\xi\cdot A} &= i\varepsilon_A' 2\left[\frac{C_3}{2}, e^{i2\xi\cdot A}\right]. \end{aligned} \tag{3.43}$$

Taking into account eqs. (3.42), (3.20), (3.26) and (3.43) we arrive at the following PCAC relations for the axial vector currents $A_\mu{}^{W,Z,\text{tot}}$

$$\partial^\mu A_\mu{}^{W,\text{tot}} = -i\frac{F^2}{4}\operatorname{Tr}\left[\begin{pmatrix} 0 & & & \\ & m_{\pi^+}^2 & & \\ & & m_{K^+}^2 & \\ & & & m_{D^0}^2 \end{pmatrix}\gamma_5\{C_+(\Theta), e^{i2\xi\cdot A}\}\right] \tag{3.44}$$

or, restricting ourselves to linear terms in ξ,

$$\partial^\mu A_\mu{}^{W,\text{tot}} = \sqrt{2}\,F\,\{\cos\Theta(m_\pi^2\text{-}\pi^- + m_F^2\text{-}F^-) + \sin\Theta(m_K^2\text{-}K^- - m_D^2\text{-}D^-)\} + O(\xi^3) \tag{3.45}$$

and, analogously,

$$\partial^\mu A_\mu{}^{Z,\text{tot}} = F\left\{m_{\pi^0}^2\pi^0 + \frac{m_{\eta_8}^2}{\sqrt{3}}\eta_8 - \sqrt{\frac{2}{3}}\,m_{\eta_{15}}^2\eta_{15}\right\} + O(\xi^3) \tag{3.46}$$

while, of course, $\partial^\mu J_\mu{}^{A,\text{tot}} = 0$.

We notice that the parameter F introduced in eq. (2.25) for dimensional reasons is now recognized as an (averaged) meson decay constant. The case of equal decay constants, $F_\pi \simeq F_K \simeq F_D \simeq F_F \cdots \simeq F$ considered here corresponds to an $SU(4)$ invariant vacuum. If necessary, F splittings can be taken into account, at least phenomenologically, by replacing in the final expressions (e.g. currents)

$$\xi_i = \Phi_i/F \to \xi_i = \Phi_i/F_i, \qquad F D_\mu \xi_i \to (\hat{F} D_\mu \xi)_i$$

etc. where $\hat{F}$ is the diagonal matrix $\hat{F} = \operatorname{diag}(F_1, \ldots, F_{15})$.

---

[9]) The expressions for $\delta W_\mu{}^i$, δX_μ follow from a transformation law analogous to (3.11) where now the derivative term must, however, be dropped (for a similar prescription applied to an Abelian gauge field see ref. [1]).

III.2.3. The effective Lagrangian

For the future practical applications to be reported in Sect. IV we need an effective Lagrangian describing weak and weak/radiative decays of ordinary and charmed hadrons to first order in the weak interaction constant $G \sim 10^{-5}/M_P^2$. Such an expression can easily be derived by considering processes with W- and Z-boson exchange in second order of the coupling constants $\varkappa$, ϱ. In a general (R_β) gauge the vector boson propagators read (cf. Appendix B)

$$D^W_{\mu\nu}(x) = \int \frac{d^4k}{(2\pi)^4} e^{-ikx} \left(-g_{\mu\nu} + \left(1 - \frac{1}{\beta}\right) \frac{k_\mu k_\nu}{k^2 - \frac{\overline{M}_W{}^2}{\beta} + i\varepsilon}\right) \frac{1}{k^2 - \overline{M}_W{}^2 + i\varepsilon}, \tag{3.47}$$

$$D^Z_{\mu\nu}(x) = \int \frac{d^4k}{(2\pi)^4} e^{-ikx} \left(-g_{\mu\nu} + \left(1 - \frac{1}{\gamma}\right) \frac{k_\mu k_\nu}{k^2 - \frac{\overline{M}_Z{}^2}{\gamma} + i\varepsilon}\right) \frac{1}{k^2 - \overline{M}_Z{}^2 + i\varepsilon}.$$

The total masses $\overline{M}_W$, $\overline{M}_Z$ of the W- and Z-bosons get contributions both from the hadron sector (see the term $\mathcal{L}_{\text{bil}}$ in Appendix C) and the Higgs mechanism of the lepton sector (see Appendix B) i.e.

$$\overline{M}_W{}^2 = (M_W{}^2)_{\text{Higgs}} + \left(\frac{F\varkappa}{\sqrt{2}}\right)^2,$$

$$\overline{M}_Z{}^2 = (M_Z{}^2)_{\text{Higgs}} + 2\left(\frac{F\varkappa}{2\cos\Theta_W}\right)^2. \tag{3.48}$$

With the usual approximations $D_{\mu\nu}(x) \sim \delta^{(4)}(x)\, g_{\mu\nu}/\overline{M}^2$, valid for large vector boson masses, one, finally, obtains

$$\mathcal{L}_{\text{eff}} = \mathcal{L}_{\text{weak}} + \mathcal{L}_{\text{weak/elm}} \tag{3.49}$$

with

$$\mathcal{L}_{\text{weak}} = -\frac{\overline{G}}{\sqrt{2}} (J^{\mu,Z} J_\mu{}^Z + J^{+\mu,W} J_\mu{}^W), \tag{3.50}$$

$$\mathcal{L}_{\text{weak/elm}} = -\frac{\overline{G}}{\sqrt{2}} e \frac{F^2}{8} A^\mu \Big\{H(C_+)\left((J_\mu{}^{0,W})^+_{\text{mes}} + \text{h.c.}\right) + \left(H\left(\frac{C_3}{2}\right) - 4\sin^2\Theta_W \langle [Q, e^{i2\xi\cdot A}]\,[Q, e^{-i2\xi\cdot A}]\rangle\right)(J_\mu{}^{0,Z})_{\text{mes}}\Big\}. \tag{3.51}$$

Here the current J_μ denotes the sum of the leptonic currents (3.41) and the "free" hadronic currents $J^0_{\mu,h} \equiv J_{\mu,h}(e = \varkappa = 0)$. $H(C)$ is a coupling function defined in Appendix C, and $\overline{G} = \sqrt{2}\,\varkappa^2/8\overline{M}_W{}^2$ is a corrected Fermi constant which coincides up to the negligible term $O\left(F^2(G/\sqrt{2})^2\right)$ with the usual expression $G = \sqrt{2}\,\varkappa^2/8(M_w{}^2)_{\text{Higgs}}$.
Notice that the effective weak Lagrangian (3.50) exhibits the usual (current x current) structure of the conventional weak interaction theory. It involves $SU(3) \times SU(3)$ [2, 14, 20] and $SU(4) \times SU(4)$ [35] substructures that have previously been derived only in a heuristic way. There is further the important fact that the weak neutral part of (3.50) does not contain strangeness changing terms like $-\overline{G}/\sqrt{2}\,(-F\partial^\mu K^0 + K^- \overleftrightarrow{\partial}^\mu \pi^+ + \cdots)$

$\times J^Z_{\mu,l}$. Thus, the decays $K_L \to \mu^+\mu^-$, $K^- \to \pi^- e^+ e^-$ etc. are forbidden, in agreement with the experimental data.
Finally, we remark that the Lagrangian $\mathscr{L}_{\text{weak/elm}}$ could also be obtained empirically from the Lagrangian $\mathscr{L}_{\text{weak}}$ by applying the minimal substitution rule for electromagnetism, i.e. $\partial_\mu \to \partial_\mu + ieA_\mu Q$, to the currents.[10])
In the subsequent Section we will mainly review the application of chiral $SU(3) \times SU(3)$ Lagrangians to the description of strong, weak and electromagnetic interactions of the pseudoscalar mesons (π, K, η, η'). Let us collect therefore some formulas specified to this case that will frequently be used in the text.
The electromagnetic interaction of the hadrons takes the form

$$\mathscr{L}_{\text{elm}} = -eA^\mu J^A_{\mu,h} \tag{3.52}$$

where

$$J^A_{\mu,h} = J^3_{\mu,h} + \frac{1}{\sqrt{3}} J^8_{\mu,h}. \tag{3.53}$$

The Lagrangian for the weak interaction of hadrons and leptons is ($\bar{G} \simeq G$)

$$\mathscr{L}^{(1)}_{\text{weak}} = -\frac{G}{\sqrt{2}} (J^+_{\mu,l} J_h{}^\mu + J^+_{\mu,h} J_l{}^\mu). \tag{3.54}$$

$J_{\mu,l}$ is the charged lepton current (comp. eq. (3.41)) (henceforth we omit the indices W, Z, A)

$$J_{\mu,l} = \bar{\nu}_{(\mu)} \gamma_\mu (1 + \gamma_5)\, \mu + \bar{\nu}_{(e)} \gamma_\mu (1 + \gamma_5)\, e \tag{3.55}$$

where μ, e, $\nu_{(\mu)}$ and $\nu_{(e)}$ are the fields of the muon, electron and the corresponding neutrinos, respectively. Finally, the $SU(3)$-part of the charged weak hadronic current is

$$J_{\mu,h} = \cos \Theta (V + A)_\mu{}^{1+i2} + \sin \Theta (V + A)_\mu{}^{4+i5}. \tag{3.56}$$

$J_{\mu,h} = J_{\mu b(q)} + J_{\mu,m}$ is a sum of baryon (quark) and meson currents. The expressions for the baryon or quark currents are[11]) (cf. eqs. (3.17), (3.19))

$$\begin{aligned} (V_b)_{i\mu} &= -i\bar{b}_k \gamma_\mu f_{ikl} b_l \\ (A_b)_{i\mu} &= -g_A \bar{b}_k \gamma_5 \gamma_\mu c_{ikl} b_l \end{aligned} \tag{3.57}$$

and

$$J_{\mu,q} = \cos \Theta \bar{u} \gamma_\mu (1 + \gamma_5)\, d + \sin \Theta \bar{u} \gamma_\mu (1 + \gamma_5)\, s. \tag{3.58}$$

Finally, to lowest order in the coupling constants $\varkappa$, ϱ the meson currents are (see Appendix D)

$$\begin{aligned} J^0_{i\mu,m} &= -F^2 D_\mu \xi_k \langle (V_i + A_i)\, A_k(\xi) \rangle \\ &= -\frac{F^2}{2} \langle (V_i + A_i)\, e^{i2\xi\cdot A}\, \partial_\mu e^{-i2\xi\cdot A} \rangle, \end{aligned} \tag{3.59}$$

where $A_k(\xi)$ was defined in eq. (3.18). Hitherto all Lagrangians and currents were treated as classical ones. To get quantized quantities, normal ordering of the relevant expressions is implicitly assumed, whenever necessary, in the following.

---

[10]) Such an independent treatment of weak and electromagnetic interactions must, however, fail in higher order processes where an internal $WW\gamma$-vertex contributes. The W-boson of the unified theory possesses an anomalous magnetic moment arising from a nonminimal term $F_{\mu\nu} W^{\mu+} W^{\nu-}$.

[11]) Note that the baryon (quark) currents contain also terms with meson fields. As these terms do not contribute to the decay processes considered in Sect. IV, they are not quoted explicitly here.

IV. Description of Strong, Electromagnetic and Weak Interactions of Mesons

Now, that we have reviewed the basic ideas on nonlinear chiral Lagrangians and collected the main formulas needed in the following, we will come to the physical applications. We shall demonstrate that a large variety of low-energy characteristics of meson physics (scattering lengths, electromagnetic radii, form factors, polarizabilities, decay rates, etc.) can be satisfactorily described in the tree and one-loop approximation on the basis of nonlinear chiral Lagrangians.

IV.1. Strong interactions: $\pi\pi$ and πK scattering

One of the first successful applications of nonlinear chiral Lagrangians to hadron physics was the evaluation of the scattering length of the $\pi\pi$ and πN systems [4—6]. Let us here demonstrate the usefulness of the nonlinear Lagrangian approach confining ourselves to $\pi\pi$ and πK scattering.
The $\pi\pi$ scattering amplitude has the form

$$\langle i_1 i_2 | S | i_3 i_4 \rangle = I + i(2\pi)^4 \, \delta^{(4)}(p_1 + p_2 - p_3 - p_4) \, [\delta_{i_1 i_2} \delta_{i_3 i_4} A(s, t, u) + \delta_{i_1 i_3} \delta_{i_2 i_4} A(t, s, u) + \delta_{i_1 i_4} \delta_{i_2 i_3} A(u, t, s)] \tag{4.1}$$

where I is the unit matrix, i_k are the pion isotopic indices, δ_{ij} denote Kronecker symbols, and $s = (p_1 + p_2)^2$, $t = (p_1 - p_3)^2$, $u = (p_1 - p_4)^2$ are usual Mandelstam variables. Taking account for the meson-meson Lagrangian $\mathcal{L}_{\pi\pi} = -\boldsymbol{\pi}^2(\partial_\mu \boldsymbol{\pi})^2/(2F)^2$ (cf. eq. (2.32)) and the meson mass term (3.21) the tree approximation leads to

$$A(s, t, u) = \frac{s - m_\pi^2}{F^2}. \tag{4.2}$$

From the Born term expression (4.2), one derives the following values for the scattering lengths of the s- and p-waves

$$a_0^0 = 0.16 m_\pi^{-1}, \qquad a_0^2 = -0.046 m_\pi^{-1}; \qquad a_1^1 = 0.03 m_\pi^{-3}, \tag{4.3}$$

which are in rough agreement with the current experimental data [36]

$$(a_0^0)^{\exp} = (0.26 \pm 0.05)\, m_\pi^{-1}, \qquad (a_0^2)^{\exp} = (-0.028 \pm 0.014)\, m_\pi^{-1}; \qquad (a_1^1)^{\exp} = (0.040 \pm 0.004)\, m_\pi^{-3}. \tag{4.4}$$

A further improvement of the results (4.3) emerges by including the one-loop contributions into the scattering amplitude [2, 37] (for discussions see below). Then[12])

$$a_0^0 = 0.19 m_\pi^{-1}, \qquad a_0^2 = -0.028 m_\pi^{-1}; \qquad a_1^1 = 0.031 m_\pi^{-3}. \tag{4.5}$$

The analogous formulas for πK-scattering are

$$\langle \pi_{i_1} \mathrm{K}_{\alpha_1} | S | \pi_{i_2} K_{\alpha_2} \rangle = I + i(2\pi)^4 \, \delta^{(4)}(p_1 + k_1 - p_2 - k_2) \times [\delta_{\alpha_1 \alpha_2} \delta_{i_1 i_2} T^{(+)} + i \varepsilon_{l i_2 i_1} \tau^l_{\alpha_2 \alpha_1} T^{(-)}]. \tag{4.6}$$

Here $\varepsilon_{l i_2 i_1}$ and $\tau^l_{\alpha_2 \alpha_1}$ are the antisymmetric tensor and Pauli matrices, respectively, and the amplitudes $T^{(\pm)}$ are related to the amplitudes of isotopic spin $I = 1/2,\ 3/2$ by

---

[12]) The results of ref. [2, 37] slightly differ from the values quoted in eq. (4.5). The reason is that in those papers a different form of the mass terms, as proposed by Gürsey [4], has been employed.

$T^{(1/2)} = T^{(+)} + 2T^{(-)}$, $T^{(3/2)} = T^{(+)} - T^{(-)}$. In the tree approximation one has

$$T^{(+)} = \frac{m_\pi^2 - p_1 p_2}{2F^2}, \qquad T^{(-)} = \frac{(k_1 + k_2)(p_1 + p_2)}{4F^2} \tag{4.7}$$

where p_i and k_i are the pion and kaon momenta. From eq. (4.7), one gets the scattering lengths

$$a^0_{1/2} = 0.14 m_\pi^{-1}, \qquad a^0_{3/2} = -0.07 m_\pi^{-1} \tag{4.8}$$

which are again in qualitative agreement with the experiment [*38*].

Thus, we have seen that the tree approximation gives quite reasonable results in the region of low energies where the terms with lowest powers in the Mandelstam variables q^2 dominate.[13]) To account for higher order power terms in q^2, one must include as a minimum, one-loop graphs, too.

Let us exemplify these ideas by considering the sacttering of massless pions. Following LEHMANN [*9b*], the next order q^2-terms may be evaluated by using eq. (4.2) and the unitarity relation. To this end, one calculates first the imaginary part of the one-loop diagram depicted in Fig. 2a where the vertices are given by the term $\mathscr{L}_{\pi\pi}$. Straightforward calculation gives

$$\operatorname{Im} A(s, t, u) = [3s^2\Theta(s) + t(t-u)\,\Theta(t) + u(u-t)\,\Theta(u)]/6\pi(2F)^4 \tag{4.9}$$

where $\Theta(s)$ is the usual step function, $\Theta(s) = 1, 0$ für $s > 0$, $s < 0$. Finally the real part $\operatorname{Re} A(s, t, u)$ can be evaluated by inserting (4.9) into the corresponding dispersion relation. This yields

$$A(s, t, u) = \frac{s}{F^2} - \frac{1}{6\pi^2(2F)^4}\{3s^2 \ln[\alpha_1(-s - i0)] + t(t-u)\ln[\alpha_2(-t - i0)] + u(u-t)\ln[\alpha_2(-u - i0)]\} + O(q^6), \tag{4.10}$$

where an arbitrary subtraction polynomial $c_1 s^2 + c_2(t-u)^2$ has been accounted for by introducing the parameters α_1, α_2. These parameters may be determined when applying the superpropagator technique [*39*].

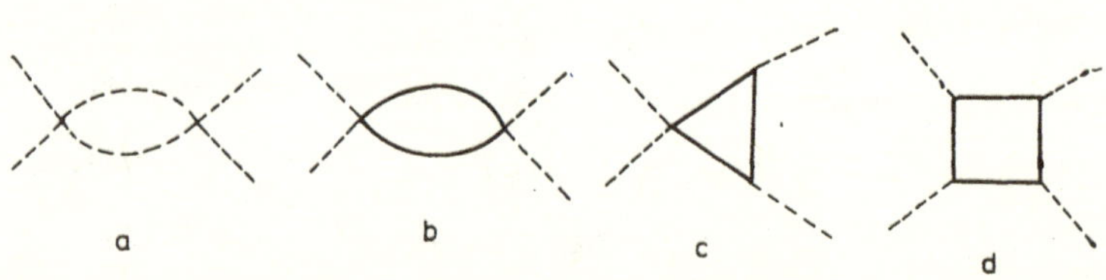

Fig. 2. Loop diagram contributing in the one-loop approximation to $\pi\pi$ scattering; ------ pion lines, ——— baryon lines

In Fig. 2 all the diagrams are depicted which contribute in the one-loop approximation. The pion loop diagram (Fig. 2a) has already been discussed above. The remaining diagrams (Fig. 2b—d) contain baryon loops. In the case of pion-nucleon interactions their vertices are defined by the Lagrangian [*9b*]

$$\mathscr{L}_{\pi N} = i\frac{M}{F} :\bar\psi\gamma_5\boldsymbol{\tau}\boldsymbol{\pi}\psi: + \frac{M}{2F^2} :\bar\psi\psi\boldsymbol{\pi}^2:. \tag{4.11}$$

On the other hand, if one wants to take into account the interaction of the pion with all the members of the $SU(3)$ baryon octet, one must choose the Lagrangians (2.34a, b)

[13]) In the cms-system we have $s = 4q^2$, $t = -2q^2(1 - \cos\vartheta)$, $u = -2q^2(1 + \cos\vartheta)$, where ϑ is the scattering angle.

[*19*][14]). We do not want to go into the details of the calculations here but only mention that the meson loop calculations may be performed with the help of the superpropagator method [*39*]. The baryon loop contributions may be evaluated by utilizing standard methods of renormalizable field theory. The reader interested in more details is referred to the works [*9b*, *19*] and [*37*, *40*] (the latter papers give also a correct treatment of the pion mass).

The papers quoted above contain, in particular, an analysis of $\pi\pi$-scattering phases. As has been found there, the energy dependence of the s- and p-wave scattering lengths is in

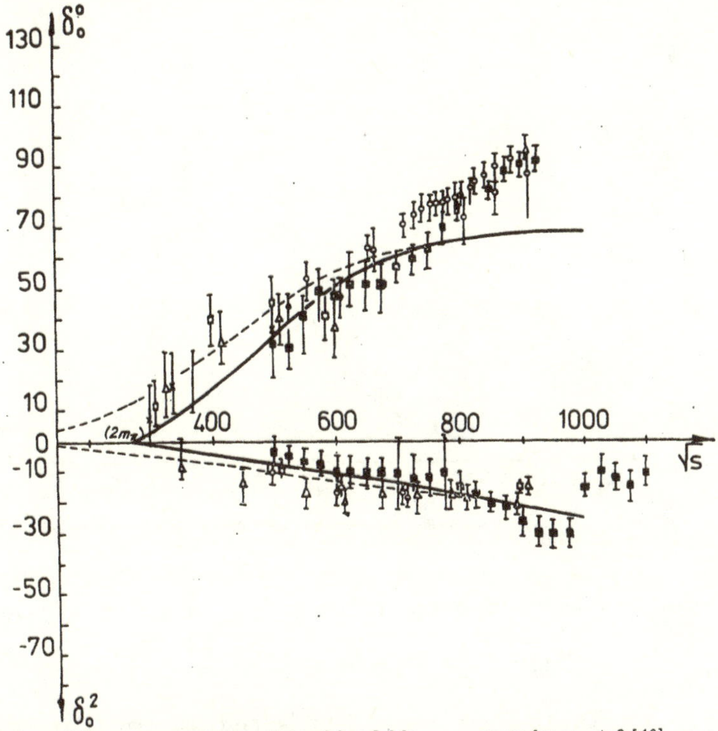

Fig. 3. The behaviour of the s-wave scattering phases $\delta_0{}^0$ and $\delta_0{}^2$; —— curve for $m_\pi \neq 0$ [*40*], ------ curve for $m_\pi = 0$ [*9b*, *19*]. The data points are taken from ref. [*41*, *42*].

the region $s \leqq 1\ \mathrm{GeV}^2$ in satisfactory agreement with the data [*41*, *42*] (cf. Figs. 3, 4). The p-wave exhibits a clear ϱ-resonance at an energy of ~ 800 MeV with a width ~ 150 MeV. Finally, as it turns out, the dominant contributions to the higher partial wave scattering lengths come from the pion loop alone. Thus far, about $\pi\pi$-scattering. In the following subsection we will describe the electromagnetic properties of mesons.

IV.2. Electromagnetic properties of mesons

IV.2.1. Pion form factor [*43*]

The matrix element for a pion in an external electromagnetic field $A_\mu(q)$ is

$$\langle \pi^+ |\ S\ | \pi^+ \rangle = iep_\mu A^\mu(q)\, F_\pi(q) . \tag{4.12}$$

[14]) In (4.11) we omitted a derivative interaction of the form (2.34c) as it contributes only insignificantly to $\pi\pi$ scattering.

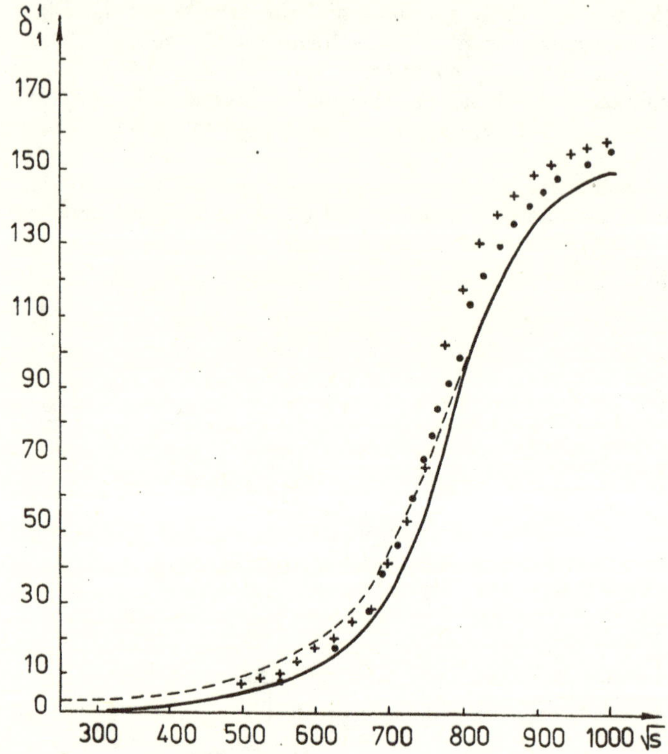

Fig. 4. The behaviour of the p-wave scattering phase δ_1^1: ——— curve for $m_\pi \neq 0$ [40], ----- curve for $m_\pi = 0$ [9b, 19]. The data points +, • are taken from ref. [41, 42].

p_2 and p_1 are the momenta of the incoming and outgoing pions, $p = p_1 + p_2$, $q = p_1 - p_2$, and

$$F_\pi(q) = 1 + F_\pi^{(\pi)}(q) + F_\pi^{(b)}(q) + \cdots \tag{4.13}$$

is the pion form factor. The first term in eq. (4.13) is the Born term (Fig. 5a), $F_\pi^{(\pi)}$ is the contribution of the pion loop (Fig. 5b) and $F_\pi^{(b)}$ is the contribution of the baryon loop diagrams (Fig. 5c—e) in the e/F^2 approximation.

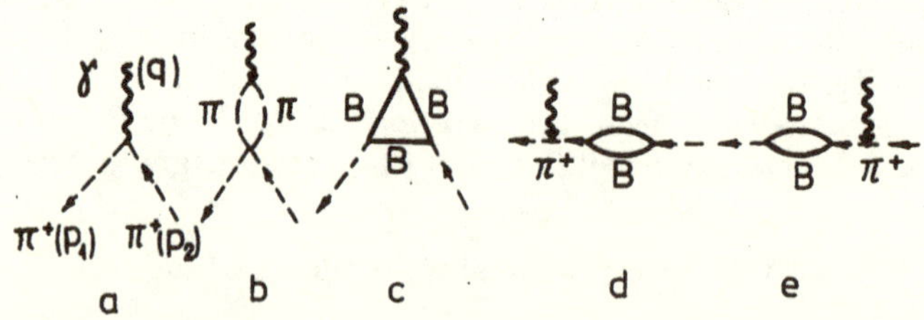

Fig. 5. Diagrams contributing to the pion form factor in the tree and one-loop approximation

The function $F_\pi^{(\pi)}(q)$ has been evaluated by the superpropagator (SP-) method giving the result

$$F_\pi^{(\pi)}(q) = \alpha_0 \left\{ \bar{q}^2 \left[\frac{13}{24} - \frac{3}{2} c + \ln \frac{2\pi F}{m_\pi} \right] - 1 + \frac{4}{3} \bar{q}^2 + (1 - \bar{q}^2)\, J(\bar{q}^2) \right\}, \tag{4.14}$$

where $\alpha_0 = 1/3\,(m_\pi/2\pi F)^2 \approx 0.02$, $\bar{q}^2 = q^2/4m_\pi^2$, $c = 0.577\ldots$ is the Euler constant, and the function $J(\bar{q}^2)$ is defined by

$$J(\bar{q}^2) = 1 - \frac{1}{2}\sum_1^\infty (4\bar{q}^2)^n \frac{n!(n-1)!}{(2n+1)!}$$

$$= \begin{cases} x \arctan x^{-1}, & x = \left(\dfrac{1}{\bar{q}^2} - 1\right)^{1/2}, \quad 0 < \bar{q}^2 < 1 \\ \dfrac{y}{2}\left[\ln\dfrac{1+y}{1-y} - i\pi\right], & y = \left(1 - \dfrac{1}{\bar{q}^2}\right)^{1/2}, \quad \bar{q}^2 > 1 \\ \dfrac{y}{2}\ln\dfrac{y+1}{y-1}, & \bar{q}^2 < 0. \end{cases}$$

From eq. (4.14) we arrive at the following pion loop contribution to the mean-square radius of the pion

$$\langle r^2\rangle_\pi^{(\pi)} = \frac{3\alpha_0}{2m_\pi^2}\left[\frac{13}{24} - \frac{3}{2}c + \ln\frac{2\pi F}{m_\pi}\right] \approx 0.06\ \text{fm}^2. \tag{4.15}$$

The baryon loop contribution has been evaluated to within an accuracy of q^2-terms only, since the terms of higher powers in q^2 are of the order $q^2 o(q^2/M^2)$ and thus nonessential at low energies. (Constant terms appearing in the computation of the baryon diagrams are mutually compensating due to the Ward identities). Then, there remain only finite integrals which can be evaluated in a standard manner yielding the result

$$F_\pi^{(b)}(q) = \frac{f_\pi(a)}{6(2\pi)^2}\, g^2 \frac{q^2}{M_N^2}. \tag{4.16}$$

The factor $f_\pi(a)$ in eq. (4.16) results from including the contributions of all the members of the baryon octet. We have

$$f_\pi(a = 2/3) = 1 + \frac{28}{27}\left(\frac{M_N}{M_{\Lambda,\Sigma}}\right)^2 + \frac{1}{9}\left(\frac{M_N}{M_\Xi}\right)^2 = 1.73. \tag{4.17}$$

In deriving this number, it was assumed that the strong coupling constants $g_{\pi\bar{B}_iB_i}$ of different baryons are approximately constant and that the only effect of symmetry breaking is reflected in the masses of the baryon propagators. Eq. (4.16) yields the following contribution to the pion mean square radius

$$\langle r^2\rangle_\pi^{(b)} = f_\pi(a)\,\frac{g^2}{(2\pi M_N)^2} \approx 0.35(\text{fm})^2. \tag{4.18}$$

By comparing eqs. (4.15), (4.18), we conclude that the dominant contribution to the pion radius comes from the baryon loop diagrams. Combining these two results, we get

$$\sqrt{\langle r^2\rangle_\pi} = \sqrt{\langle r^2\rangle_\pi^{(\pi)} + \langle r^2\rangle_\pi^{(b)}} \approx 0.64\ \text{fm},\ \langle r^2\rangle_\pi = 0.41\ \text{fm}^2, \tag{4.19}$$

a value which is close to the results obtained in the models with ϱ-dominance $\left(\sqrt{\langle r^2\rangle_\pi} \approx \sqrt{6/m_\varrho^2} \approx 0.62\ \text{fm}\right)$.

It is instructive to recalculate once more the pion radius in the framework of the chiral meson-quark Lagrangian (2.37). Straightforward calculation of the quark loop diagrams gives

$$\langle r^2\rangle_\pi^{(q)} = 3\,\frac{g_A'^2}{(2\pi F)^2} \tag{4.18'}$$

where, in order to avoid troubles with the nonappearance of quarks, quark masses have been assumed to be very heavy, say $\gg 1$ GeV. Comparing (4.18') with (4.18) and making use of the Goldberger-Treiman relation $g^2/M_N^2 \approx g_A^2/F^2$, we see that the (SU_3) factor $f_\pi g_A^2$ is now replaced by the (color) factor $3g_A'^2$. Taking $g_A' \approx 1$ (see footnote on page 75) one, finally, obtains $\langle r^2\rangle_\pi^{(q)} \approx 0.33$ fm^2 [44] which is close to the value in eq. (4.18).

Comparing the above theoretical predictions with the data, we remind the reader that the present experimental situation is somewhat ambiguous. Different types of experiments give different results: 1) high energy scattering of pions on atomic electrons yields $\langle r^2\rangle_\pi^{\exp} = (0.31 \pm 0.04)$ fm^2 [45], 2) from threshold reactions with pion generation, $eN \to e\pi N$, one gets $\sqrt{\langle r^2\rangle_\pi^{\exp}} = (0.45 \pm 0.10)$ fm [46], 3) finally, the reaction $e^+e^- \to \pi^+\pi^-$ yields the largest value $\langle r^2\rangle_\pi^{\exp} = (0.460 \pm 0.011)$ fm^2 [47] when compared with the two preceding cases. As we see, our result (4.19) is closest to the value obtained in the latter experiment.

Finally, inserting the functions (4.14), (4.16) into (4.13) we arrive at the following expression of the form factor

$$F_\pi(q) = 1 + \alpha_0\{-1 + 9\bar{q}^2 + (1 - \bar{q}^2)\,J(\bar{q}^2)\}\,. \tag{4.20}$$

This formula describes the q^2-dependence of the pion form factor for momenta $|q^2| < 1$ GeV^2 in good agreement with the experimental data obtained at Dubna and Serpukhov [48, 49] (comp. Figs. 6a, b. The data points ⌶ are from ref. [48], ⌽ from ref. [49]).

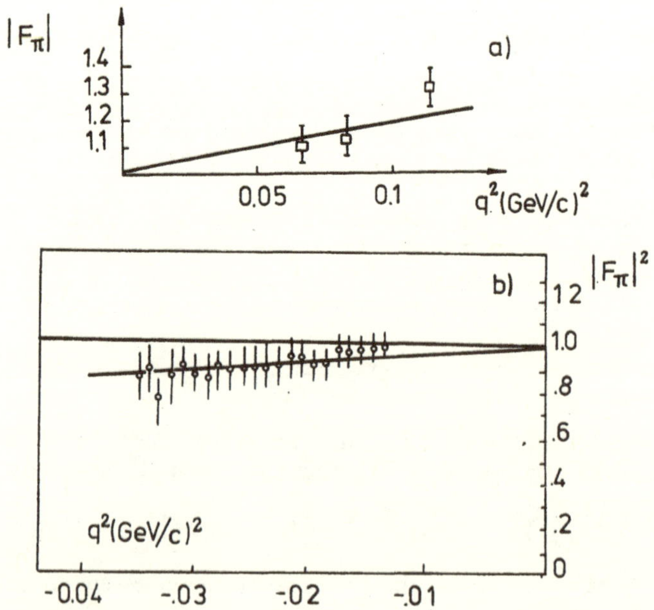

Fig. 6. The behaviour of the pion-form factor for a) $q^2 > 0$ and b) $q^2 < 0$; —— curve from ref. [37, 43]. The data points ⌶ and ⌽ are taken from ref. [48, 49]

IV.2.2. Compton effect and pion polárizability [50]

The matrix element for the Compton effect on pions is defined by

$$\langle \pi^a(p_1)\, \pi^b(p_2) |\, S\, | \gamma_{\lambda_1}(q_1)\, \gamma_{\lambda_2}(q_2) \rangle = (2\pi)^4\, i\delta^{(4)}(p_1 + p_2 - q_1 - q_2)\, \varepsilon^{\mu}_{\lambda_1} \varepsilon^{\nu}_{\lambda_2}\, T^{\mu\nu}_{ab}(p_1 p_2 \mid q_1 q_2) \tag{4.21}$$

where q_1, q_2 and $\varepsilon^{\mu}_{\lambda_1}$, $\varepsilon^{\nu}_{\lambda_2}$ are the momenta and polarization vectors of the photons, p_1 and p_2 are the momenta of the pions, and a, b are their isotopic indices.
Notice that, within the one-loop approximation, in this process the divergences cancel not only in baryon loops but also in pion loops, so that there is no need of using the SP-technique. Confining ourselves to terms in second order of the electromagnetic coupling constant e, we get the following expression for the scattering amplitude

$$\begin{aligned} T^{\mu\nu}_{ab} = {} & 2e^2(\delta_{ab} - \delta_{3a}\delta_{3b}) \left\{ g^{\mu\nu} - \frac{p_1{}^{\mu} p_2{}^{\nu}}{p_1 q_1} - \frac{p_1{}^{\nu} p_2{}^{\mu}}{p_1 q_2} \right. \\ & \left. + (g^{\mu\nu} q_1 q_2 - q_1{}^{\nu} q_2{}^{\mu})\, [\beta_{\pi}{}^{(\pi)}(q_1 q_2) + \beta_{\pi}{}^{(N)}(q_1 q_2)] \right\} \\ & + 4e^2 \delta_{3a}\delta_{3b}(g^{\mu\nu} q_1 q_2 - q_1{}^{\nu} q_2{}^{\mu})\, \beta_{\pi}{}^{(\pi)}(q_1 q_2) . \end{aligned} \tag{4.22}$$

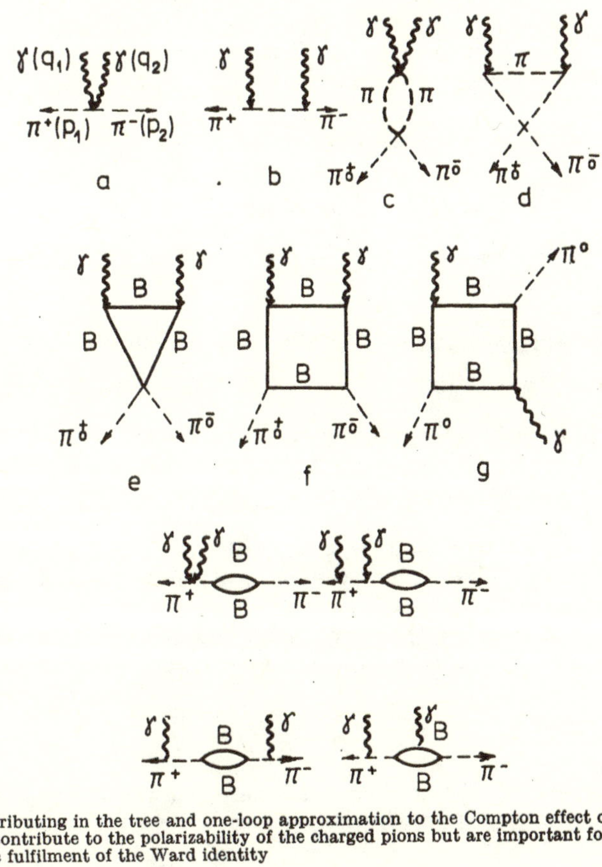

Fig. 7. Diagrams contributing in the tree and one-loop approximation to the Compton effect on pions. The last four diagrams do not contribute to the polarizability of the charged pions but are important for the cancellation of divergences and the fulfilment of the Ward identity

Here the first three terms in the bracket correspond to the Born terms (Fig. 7a and b); $\beta_\pi^{(\pi)}$ and $\beta_\pi^{(N)}$ are the contributions to the amplitude arising from the pion (Fig. 7c, d) and nucleon (Fig. 7e, f, g) loops, respectively. In $\beta_\pi^{(N)}$ only the constant terms will be kept, since the remaining ones go like $o(m_\pi^2/M^2)$ or $o(q_1q_2/M^2)$ and thus are small in the energy range of question. Besides, in deriving eq. (4.22) the relations $(q_1\varepsilon_1) = (q_2\varepsilon_2) = 0$, $q_1^2 = q_2^2 = 0$, $p_1^2 = p_2^2 = m_\pi^2$ have been used. Combining the contributions of the pion loops depicted in Figs. 7c, d we arrive at the final expression

$$\beta_\pi^{(\pi)}(q_1q_2) = (4\pi F)^{-2}\left(\frac{q_1q_2}{2m_\pi^2} - \frac{1}{3}\right) I\left(\frac{q_1q_2}{2m_\pi^2}\right),$$
$$I(\xi) = \frac{1}{\xi}\left\{\frac{1}{\xi}\left[\arctan\left(\frac{1}{\xi} - 1\right)^{-1/2}\right]^2 - 1\right\}. \tag{4.23}$$

Note that the consideration of kaon loops gives a negligible correction of order $o(m_\pi^2/m_K^2)$ to $\beta_\pi(q_1q_2)$. Let us now come to the baryon loops. When considering the Compton effect by a neutral pion one finds that all the contributions from the diagrams 7e, f, g cancel. On the other hand, for charged pions the nucleon loop diagrams 7e, f give the nonvanishing contribution

$$\beta_\pi^{(N)} = \frac{2}{3}\frac{g_A^2}{(4\pi F)^2}. \tag{4.24}$$

To account for the other members of the baryon octet, again a factor $f_\pi(a) \approx 1.7$ must be included.

The pion polarizability is defined as the coefficient appearing in the effective interaction of a pion with an external electromagnetic field,[15])

$$V_{\text{int}} = -\frac{\alpha_\pi}{2}(\boldsymbol{E}^2 - \boldsymbol{H}^2). \tag{4.25}$$

Then, taking into account that $I(0) = 1/3$, we get

$$\alpha_{\pi^\pm} = \alpha_{\pi^\pm}(q_1q_2)|_{q_1q_2=0} = \left(\beta_\pi^{(\pi)}(0) + \beta_\pi^{(b)}\right) e^2/m_\pi = 0.29\alpha/m_\pi^3 \approx 6 \cdot 10^{-3}\,\text{fm}^3 \tag{4.26}$$

$$\alpha_{\pi^0} = \alpha_{\pi^0}(q_1q_2)|_{q_1q_2=0} = 2\beta_\pi^{(\pi)}(0)\, e^2/m_\pi = -0.036\alpha/m_\pi^3$$
$$(\alpha = e^2/4\pi = 1/137) \qquad = -7.3 \cdot 10^{-4}\,\text{fm}^3. \tag{4.27}$$

The function $\beta_\pi^{(\pi)}(q_1q_2)$ is rapidly varying in the threshold region. At the two-pion threshold we have

$$\alpha_{\pi^\pm}(2m_\pi^2) = 0.5\alpha/m_\pi^3, \qquad \alpha_{\pi^0}(2m_\pi^2) = 0.3\alpha/m_\pi^3. \tag{4.28}$$

The values of $\alpha_{\pi^\pm}$ quoted above coincide in the order of magnitude with the estimates based on current algebra [*51*] but differ by a factor of 2 from them. In contrast with ref. [*51*], one has further $\alpha_{\pi^0} \neq 0$.

It is again instructive to compare the baryon loop contribution (4.24) with an analogous contribution arising from quark loops [*52*]. Straightforward calculation gives

$$\beta_{\pi^\pm}^{(q)} = \frac{2}{3}\frac{3}{(4\pi F)^2}. \tag{4.24'}$$

---

[15]) Note that the factor $g^{\mu\nu}q_1q_2 - q_1^\mu q_2^\nu$ in eq. (4.22) corresponds to the combination $(\boldsymbol{E}^2 - \boldsymbol{H}^2)$. Hence the electric and magnetic polarizabilities of the pions are equal in magnitude but opposite in sign.

Comparing eq. (4.24) (with a factor $f_\pi(a)$ included) and (4.24') one concludes that substitution of baryons by quarks simply means substitution of the baryon factor $g_A{}^2 f_\pi(a)$ by the color factor 3. Moreover, in the case of neutral pions we get again $\beta_{\pi^0}^{(q)} = 0$ due to a cancellation of the corresponding triangle and box quark loops, in complete parallel with the results of the meson-baryon model.

IV.2.3. Kaon form factor and polarizability [53]

By analogy with the pion case the form factor of the charged kaon can be written in the form

$$F_K(q) = 1 + F_K{}^{(\pi)}(q) + F_K{}^{(K)}(q) + F_K{}^{(b)}(q) + \cdots \tag{4.29}$$

where the terms in the r.h.s. of (4.29) denote the Born term and the contributions to the form factor from the pion, kaon and baryon loops in the e/F^2 approximation, respectively. Like for the pion form factor, the contribution from the kaon loop can be neglected. Let us quote therefore only the expressions for $F_K{}^{(\pi)}(q)$ and $F_K{}^{(b)}(q)$. We have

$$F_K{}^{(\pi)}(q) = \alpha_0 \left\{ \frac{q^2}{8m_\pi{}^2} \left[\ln \left(2 \left(\frac{2\pi F}{m_\pi} \right)^2 \right) - 3c + 1 \right] - 1 + \frac{q^2}{3m_\pi{}^2} + \left(1 - \frac{q^2}{4m_\pi{}^2} \right) J \left(\frac{q^2}{4m_\pi{}^2} \right) \right\}. \tag{4.30}$$

The constants α_0, c and the function $J(q^2/4m_\pi{}^2)$ are the same as in formula (4.14). Eq. (4.30) contributes to the kaon mean square radius as follows:

$$\langle r^2 \rangle_{K^\pm}^{(\pi)} = 0.08 \text{ fm}^2. \tag{4.31}$$

Again the contribution from baryon diagrams turns out to be essentially larger and equals

$$F_K{}^{(b)}(q) = \frac{f_K(a)}{6(2\pi)^2} g^2 \frac{q^2}{M_N{}^2},$$

where, as usual, a factor $f_K(a = 2/3) = 1.4$ accounts for the $SU(3)$ baryon intermediate states. Hence, it follows that

$$\langle r^2 \rangle_{K^\pm}^{(b)} = f_K(a) \frac{g^2}{(2\pi M_N)^2} \approx 0.29 \text{ fm}^2. \tag{4.32}$$

Combining eqs. (4.31) and (4.32), we get the following value for the charged kaon radius

$$\sqrt{\langle r^2 \rangle_{K^\pm}} \approx 0.6 \text{ fm}, \qquad \langle r^2 \rangle_{K^\pm} = 0.37 \text{ fm}^2. \tag{4.33}$$

For neutral kaons the baryon loop contribution is vanishing while the contributions of the pion loops are equal in magnitude but opposite in sign to those of the charged case. Thus, we have

$$\langle r^2 \rangle_{K^0} = -\langle r^2 \rangle_{K^\pm}^{(\pi)} = -0.08 \text{ fm}^2. \tag{4.34}$$

For comparison, we quote again the value of the kaon mean square radius evaluated from quark-loop diagrams [44]

$$\langle r^2 \rangle_{K^\pm} = 0.4 \text{ fm}^2, \qquad \langle r^2 \rangle_{K^0} = -0.05 \text{ fm}^2.$$

A similar value for the neutral kaon radius has also been derived within vector dominance models with current mixing [54]. Let us quote the latest experimental data for $\langle r^2 \rangle_K$

extracted from Ke-scattering

$$\langle r^2\rangle^{\text{exp}}_{K^\pm} = (0.26 \pm 0.07)\ \text{fm}^2 \quad [55]$$
$$\langle r^2\rangle^{\text{exp}}_{K^0} = (-0.054 \pm 0.026)\ \text{fm}^2 \quad [56]. \tag{4.35}$$

In conclusion, we briefly discuss the Compton effect on kaons. The amplitudes with charged and neutral external kaons take the form (the Born term is omitted)

$$T_+^{\mu\nu} = 2e^2(g^{\mu\nu}q_1q_2 - q_1^{\nu}q_2^{\mu})\,[\beta_K^{(\pi)}(q_1q_2) + \beta^{(K)}_{K^+}(q_1q_2) + \beta^{(b)}_{K^+}]$$
$$T_0^{\mu\nu} = 2e^2(g^{\mu\nu}q_1q_2 - q_1^{\nu}q_2^{\mu})\,[\beta_K^{(\pi)}(q_1q_2) + \beta^{(K)}_{K^0}(q_1q_2)] \tag{4.36}$$

where in the e^2/F^2 approximation

$$\beta_K^{(\pi)}(q_1q_2) = (4\pi F)^{-2}\,\frac{q_1q_2}{4m_\pi^2}\,I\left(\frac{q_1q_2}{2m_\pi^2}\right),$$
$$\beta^{(K)}_{K^+}(q_1q_2) = (8\pi F)^{-2}\left(1 + \frac{q_1q_2}{2m_K^2}\right) I\left(\frac{q_1q_2}{2m_K^3}\right),$$
$$\beta^{(K)}_{K^0}(q_1q_2) = (4\pi F)^{-2}\,\frac{q_1q_2}{4m_K^2}\,I\left(\frac{q_1q_2}{2m_K^3}\right), \tag{4.37}$$
$$\beta^{(b)}_{K^+}(q_1q_2) = \frac{2}{3}\,\frac{f_K(a)\,g^2}{(4\pi M_N)^2} \approx 1.4(4\pi F)^{-2},$$

the function $I(\xi)$ being defined in eq. (4.23). Notice that for neutral external kaons (like for neutral external pions) the baryon loop diagrams do not contribute to $T_0^{\mu\nu}$. It follows from eq. (4.37) that at $q_1q_2 = 0$ the main contribution to the amplitude $T_+^{\mu\nu}$ comes from the baryon diagrams, whereas $T_0^{\mu\nu}(0) = 0$. Defining the kaon polarizatility by formulas analogous to eqs. (4.26), (4.27), one, finally, has

$$\alpha_{K^\pm} = \frac{1.5}{m_K}\left(\frac{e}{4\pi F}\right)^2 \approx 1.6 \cdot 10^{-3}\ \text{fm}^3$$
$$\alpha_{K^0} = 0. \tag{4.38}$$

These values are consistent with theoretical estimates derived on the basis of current algebra and PCAC [57], $\alpha_{K^+} \sim 10^{-3}\ \text{fm}^3$, and with the experimental data $\alpha^{\text{exp}}_{K^+} \sim -(4 \pm 11) \cdot 10^{-3}\ \text{fm}^3$ [58].

IV.3. The main decay modes of the $SU(3)$ mesons (π, K, η, η')

As another illustration of physical applications of chiral Lagrangians, consider leptonic, semileptonic and radiative decays of pseudoscalar mesons. In the one-loop approximation the dominant contributions to the decay amplitudes turn out to arise again from baryon loop diagrams. Contributions from pion loops are either numerically small or even zero. It is therefore sufficient to investigate only tree and baryon loop diagrams in the following.

As already mentioned, the divergences of baryon loop diagrams may be treated by standard methods of renormalizable field theories. Besides, in many cases the divergences of different baryon loop graphs turn out to cancel each other as a result of the underlying symmetry properties of the theory. Thus, special regularization techniques going beyong ordinary field theory are not required. We will now demonstrate how a rather complete picture of the decays of the $SU(3)$ pseudoscalar mesons (π, K, η, η') emerges on the basis of the Lagrangian of Sect. III.

IV.3.1. Pion decays

Let us consider as a first example pion decays. The relevant part of the Lagrangian describing weak and electromagnetic interactions of pions and nucleons with leptons takes the form

$$\mathcal{L}^{(\pi N)}_{\text{weak/elm}} = G \cos \Theta J^{+}_{\mu,l} : \Big[F(\partial^\mu \pi^- - ieA^\mu \pi^-) + i(\pi^- \partial^\mu \pi^0 - \pi^0 \partial^\mu \pi^-) - \frac{1}{\sqrt{2}} \bar{\Psi}_P \gamma^\mu (1 + g_A \gamma_5) \Psi_N \Big] : \tag{4.39}$$

where the lepton current $J_{\mu,l}$ is given by eq. (3.55).

IV.3.1a. Leptonic decays: $\pi^\pm \to \mu^\pm \nu_\mu (e^\pm \nu_e)$ [59]

Let us start with the weak leptonic decay of the pion, $\pi^\pm \to \mu^\pm \nu_\mu (e^\pm \nu_e)$ which is usually taken to fix the dimensional parameter F. In the tree approximation (see Fig. 8a), one has[16]

$$T^{(a)}_{\pi \to \mu\nu_\mu} = i \frac{G}{\sqrt{2}} \cos \Theta \left(\sqrt{2}\, F p^\mu l_\mu^{(+)} \right), \qquad l_\mu^{(+)} = \bar{u}_{(\mu)} \gamma_\mu (1 + \gamma_5)\, u_{(\nu)}, \tag{4.40}$$

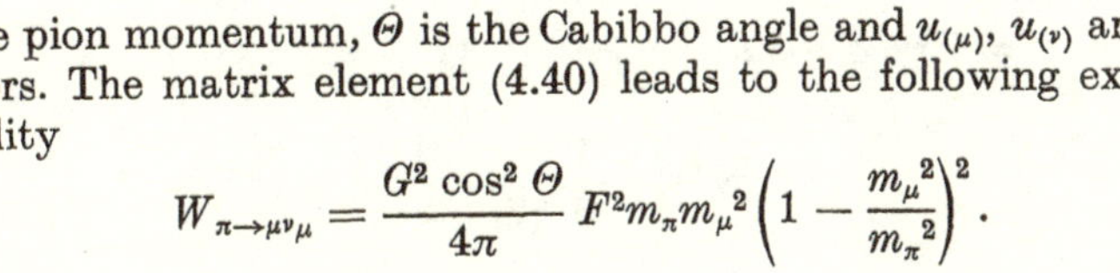

Fig. 8. Tree and baryon loop diagrams describing the process $\pi^+ \to \mu^+ \nu_\mu$

where p_μ is the pion momentum, Θ is the Cabibbo angle and $u_{(\mu)}$, $u_{(\nu)}$ are the muon and neutrino spinors. The matrix element (4.40) leads to the following expression for the decay probability

$$W_{\pi \to \mu\nu_\mu} = \frac{G^2 \cos^2 \Theta}{4\pi} F^2 m_\pi m_\mu^2 \left(1 - \frac{m_\mu^2}{m_\pi^2}\right)^2. \tag{4.41}$$

Comparing the theoretical estimate for $W_{(\pi \to \mu\nu_\mu)}$ with the experimental value [60]

$$W^{\text{exp}}_{(\pi \to \mu\nu_\mu)} = 3.84 \cdot 10^7\ s^{-1}$$

one finds

$$F = 95\ \text{MeV}. \tag{4.42}$$

[16]) Note that if $m_\pi = 0$ the decay $\pi \to \mu\nu_\mu$ is completely determined by the tree graphs. This is so because the contributions of all the other diagrams cancel thus giving zero.

Let us next turn to the one-loop diagrams shown in Fig. 8b, c. The net effect of these graphs will be a renormalization of the pion mass, wave function and decay constant F, respectively. Consider the two diagrams exhibited in Fig. 8b, c. Their matrix elements are

$$T^{(b)} = T^{(a)} \frac{\Pi(p^2)}{m_\pi^2 - p^2}, \tag{4.43}$$

$$\begin{aligned} T^{(c)} &= -l_\mu^{(+)} G \cos \Theta \frac{g_A M}{F(2\pi)^4} \int d^4k \, \mathrm{Tr}\{\gamma_5 (M - \hat{k})^{-1} \gamma_\mu \gamma_5 (M - \hat{k} - \hat{p})^{-1}\} \\ &= T^{(a)} R(m_\pi^2), \qquad (\hat{k} = \gamma_\mu k^\mu). \end{aligned} \tag{4.44}$$

The polarization operator $\Pi(p^2)$ and the function $R(p^2)$ are defined by

$$\begin{aligned} \Pi(p^2) &= -i \frac{2g^2}{(2\pi)^4} \int d^4k \, \mathrm{Tr}\,\{\gamma_5 (M - \hat{k})^{-1} \gamma_5 (M - \hat{k} - \hat{p})^{-1}\} = \mathrm{const} + p^2 R(p^2) \\ &\equiv \Pi(m_\pi^2) - \frac{\partial}{\partial p^2} \Pi(p^2)_{p^2 = m_\pi^2} (m_\pi^2 - p^2) + \cdots, \\ R(p^2) &= -i \left(\frac{g}{2\pi^2}\right)^2 \int \frac{d^4k}{(M^2 - k^2 - i\varepsilon)\left(M^2 - (k - p)^2 - i\varepsilon\right)}. \end{aligned} \tag{4.45}$$

As we find, the self-energy insertion into the pion line of diagram 8b leads to a renormalization of the pion mass and wave function. Indeed, adding $T^{(a)}$ and $T^{(b)}$, we get (for the moment, we include a pion propagator to have Green functions)

$$T^{(a)} \left(\frac{1}{m_\pi^2 - p^2} + \frac{\Pi(p^2)}{(m_\pi^2 - p^2)^2}\right) \approx T^{(a)} \frac{Z_\pi}{\overline{m}_\pi^2 - p^2},$$

where

$$\begin{aligned} \overline{m}_\pi^2 &= m_\pi^2 - \Pi(m_\pi^2) \\ Z_\pi &= (1 + \Pi'(p^2)|_{p^2 = m_\pi^2})^{-1}. \end{aligned} \tag{4.46}$$

Thus, at the new mass shell $p^2 = \overline{m}_\pi^2 = m_\pi^2 - \Pi(m_\pi^2)$ the amplitude takes the value $Z_\pi T^{(a)}(\overline{m}_\pi^2)$. Finally, including also the third diagram 8c, the total amplitude is

$$T = T^{(a)}[Z_\pi + R(m_\pi^2)] \approx T^{(a)}[1 - m_\pi^2 R'(m_\pi^2)],$$

$$R'(m_\pi^2) = \frac{1}{6} \left(\frac{g_A}{2\pi F}\right)^2.$$

This leads altogether to a renormalization of the decay constant F,

$$F \to F \left[1 - \frac{1}{6} \left(\frac{g_A m_\pi}{2\pi F}\right)^2\right] = F(m_\pi^2) = 95 \text{ MeV}, \tag{4.47}$$

and thus

$$F = 96 \text{ MeV}.$$

Clearly, this correction in F can be safely omitted in practical applications.

IV.3.1b Semileptonic Decay: $\pi^+ \to \pi^0 e^+ \nu_e$

As a second example we consider the β-decay of the charged pion, $\pi^+ \to \pi^0 e^+ \nu_e$ (Fig. 9). The evaluation of the decay amplitude runs in parallel to the calculation of the electromagnetic form factor [43] and results in the expression

$$T_{(\pi^+\to\pi^0 e^+\nu_e)} = T^{(0)} \left[1 + \frac{1}{6}\left(\frac{g_A}{2\pi F}\right)^2 q^2\right] + \frac{1}{6}\, G \cos \Theta \left(\frac{g_A}{2\pi F}\right)^2 (m_{\pi^0}^2 - m_{\pi^+}^2)\, q^\nu l_\nu^{(+)} \qquad (4.48)$$

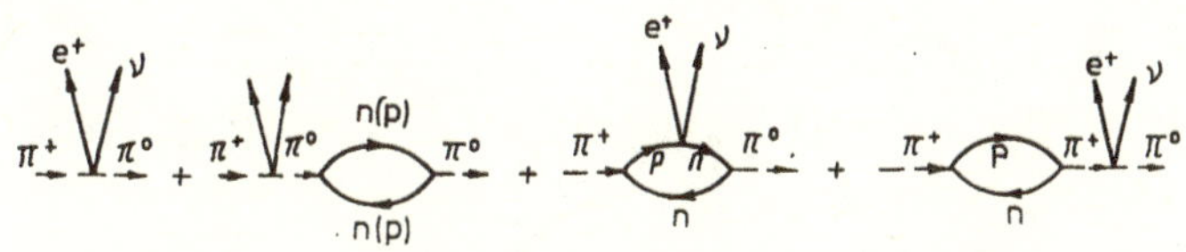

Fig. 9. Tree and baryon loop diagrams contributing to the beta-decay of π^+

where

$$T^{(0)} = G \cos \Theta (p_{\pi^+} + p_{\pi^0})^\mu\, l_\mu^{(+)}, \qquad q = p_{\pi^+} - p_{\pi^0}$$

is the Born amplitude. The last term in the r.h.s. of eq. (4.48) is of an order of $0(m_e/m_\pi)$ and therefore negligibly small. It is worth remarking that the weak radius of the pion following from eq. (4.48) coincides in magnitude with the electromagnetic radius (cf. eq. (4.18); as usual, the $SU(3)$-factor $f_\pi(a) = 1.7$ must be included).

IV.3.1c. Radiative decays: $\pi^0 \to 2\gamma$, $\pi^+ \to \mu^+ \nu_\mu \gamma$

The amplitude of the decay $\pi^0 \to 2\gamma$ has been investigated a long time ago by Steinberger [61] who found apparent ambiguities in the evaluation of the associated diagram shown in Fig. 10. The reason for such ambiguities is an insufficient convergence of

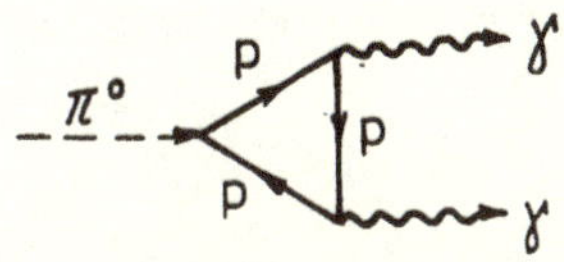

Fig. 10. Baryon loop diagram describing the decay $\pi^0 \to 2\gamma$

the loop integral of the "anomalous" triangle diagram containing two vector and one axial vector vertex. Since the above decay seemingly contradicts the standard approach of current algebra, it has attracted a great deal of attention in the literature [10]. Later on it was proved that it can be reconciled with current algebra if the usual PCAC condition for the currents (cf. eqs. (3.44, 45)) are modified by including so-called anomalous terms. Note, that in the case where anomalous diagrams are present, they determine almost completely the amplitudes of the respective decay processes.
After these general remarks, let us quote the explicit expression for the amplitude $\pi^0 \to 2\gamma$. We have

$$T_{\mu\nu}^{(\pi^0)} = -f_{(\pi^0)} \varepsilon_{\mu\nu\alpha\beta} q_1{}^\alpha q_2{}^\beta$$

with

$$f_{(\pi^0)} = \frac{e^2 g_A}{(2\pi)^2 F} = \frac{g_A m_\pi \alpha}{F \pi m_\pi} \approx 0.56\alpha/m_\pi, \qquad (4.49)$$

where q_i are photon momenta and $\varepsilon_{\pi\nu\alpha\beta}$ is the fully antisymmetric tensor. The experimental value $f^{\exp}_{(\pi^0)}$ equals

$$f^{\exp}_{(\pi^0)} \approx 0.47\alpha/m_\pi . \tag{4.50}$$

$$\pi^+ \to \mu^+\nu_\mu\gamma$$

Let us now turn briefly to the process $\pi^+ \to \mu^+\nu_\mu\gamma$. For a more detailed discussion we refer the reader to the literature [*59*].
The Born approximation is defined by the diagrams shown in Fig. 11

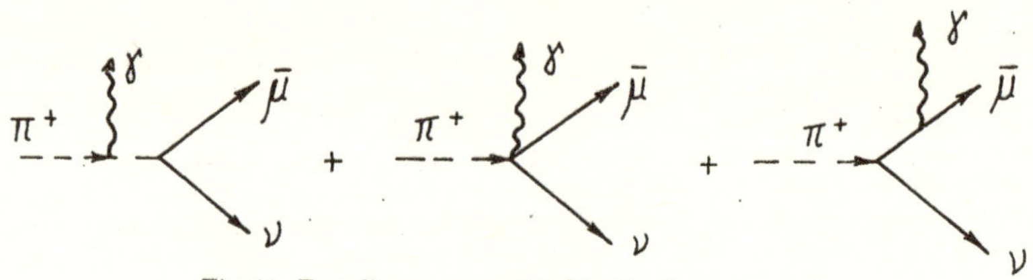

Fig. 11. Tree diagrams associated to the decay $\pi^+ \to \mu^+\nu_\mu\gamma$

$$T_\mu{}^{(0)} = i \cos \Theta eGF \left[g_{\mu\nu} + \frac{p_\mu(p-q)_\nu}{pq} \right] l^{\nu(+)}$$
$$- i \cos \Theta eGF \bar{u}_{(\mu)}\gamma_\mu(\hat{k} + \hat{q} - m_\mu)^{-1} \hat{p}(1+\gamma_5)\, u_{(\nu)} . \tag{4.51}$$

where p, k and q are the momenta of the pion, muon and photon, respectively, and m_μ is the muon mass. Again we are mainly interested in the one-loop diagrams depicted in Figs. 12, 13. (The arrows indicate the lines emitting a photon. Fig. 12 is therefore a short-hand notation for five diagrams, Fig. 13 for three diagrams). They yield the contributions

$$T_\mu{}^{(1)} = \left[-R(m_\pi{}^2) - \frac{1}{6} \left(\frac{g_A m_\pi}{2\pi F} \right)^2 \right] T_\mu{}^{(0)},$$
$$T_\mu{}^{(2)} = R(m_\pi{}^2)\, T_\mu{}^{(0)} - i \cos \Theta eG[ih_V \varepsilon_{\mu\nu\alpha\beta} p^\alpha q^\beta - h_A(g_{\mu\nu}pq - p_\mu q_\nu)]\, l^{\nu(+)}, \tag{4.52}$$

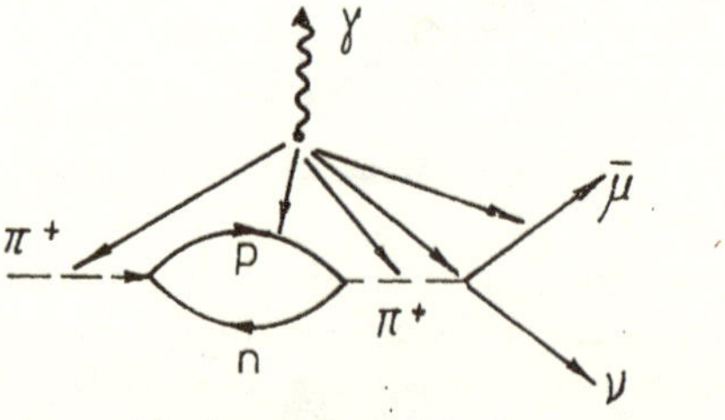

Fig 12. Five baryon loop diagrams of the type shown in Fig. 8b contributing to the decay $\pi^+ \to \mu^+\nu_\mu\gamma$

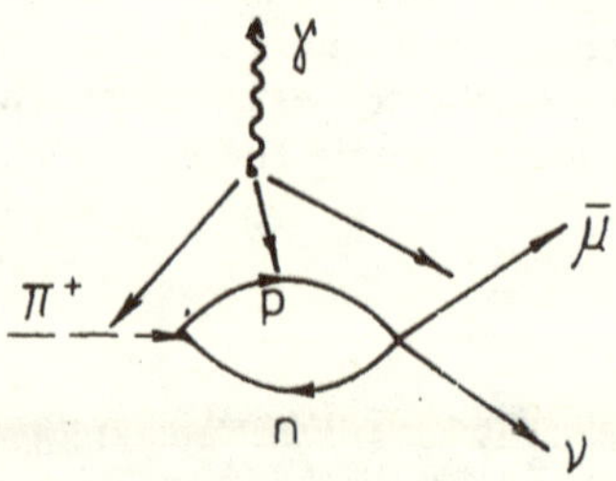

Fig. 13. Three baryon loop diagrams of the type shown in Fig. 8c contributing to the decay $\pi^+ \to \mu^+\nu_\mu\gamma$

where

$$h_V = g_A/8\pi^2F, \qquad h_A = g_A^2/6(2\pi)^2\, F.$$

This way, the effect of the nucleon diagrams is first a renormalization of the decay constant F, and second the appearance of structure constants in the emission process. Notice that the ratio $\gamma = h_A/h_V$ takes the value

$$\gamma = \frac{1}{3}\, g_A \approx 0.42$$

which is in good agreement with recent experimental data, $\gamma^{\exp} \approx 0.44 \pm 0.12$ [63]. Moreover, it is worth mentioning that the one-loop approximation respects the following current algebra relations between $f_{(\pi^0)}$ (cf. eq. (4.49)) and h_V, and β_π (cf. eq. (4.24)) and h_A [51]

$$h_V = f_{(\pi^0)}/2e^2, \qquad h_A = F\beta_\pi^{(N)}.$$

Up to now we have considered only nucleon loop diagrams. The inclusion of the other members of the baryon octet leads to the following modifications in the expressions for the $\pi^+ \to \pi^0 e^+ \nu_e$, $\pi^0 \to 2\gamma$ and $\pi^+ \to \mu^+ \nu_\mu \gamma$ amplitudes

$$g_A^2 \to f_\pi(a)\, g_A^2 \approx 1.7 g_A^2$$

$$g_A \to \left[1 + \frac{1}{3}\left(\frac{M_N}{M_\Xi}\right)\right] g_A \approx 1.24 g_A.$$

This yields

$$f_{(\pi^0)} \approx 0.7\alpha/m_\pi, \qquad \gamma \approx 0.5. \tag{4.53}$$

Note that the relations between h_V and $f_{(\pi^0)}$ and h_A, $\beta_\pi^{(N)}$ remain unchanged.
Thus far, about pion decays. In the above applications we confined ourselves to the most characteristic decay modes of the pion to demonstrate how the chiral Lagrangians of the strong and weak interactions do work. Thereby, we did not study such decay modes as $\pi^0 \to e^+e^-\gamma$ or $\pi^+ \to e^\pm \nu_e e^+ e^-$, since they may be traced back to the decays $\pi^0 \to 2\gamma$, $\pi^\pm \to e^\pm \nu_e \gamma$ studied above by replacing the e^+e^- pair by a photon [64]. Let us next proceed to the investigation of the η, η' decays.

IV.3.2. Radiative decays of eta-mesons

Quite recently, there appeared new experimental data for the decay $\eta' \to 2\gamma$ [65] and the mixing angle φ of singlet-octet mixing of eta-mesons [66]. The angle φ is defined by

$$\eta = \eta_8 \cos\varphi - \eta_0 \sin\varphi$$

$$\eta' = \eta_8 \sin\varphi + \eta_0 \cos\varphi.$$

At present, two values for φ are known: there is the commonly accepted value $\varphi = -11°$ [67] and the recent value $\varphi \approx -18°$ [66, 68].
In this Section we will review the radiative decays $(\eta, \eta') \to 2\gamma$ and $(\eta, \eta') \to \pi^+\pi^-\gamma$ in the light of the new measurements mentioned above. Moreover, we will add a short discussion of the decay $\eta \to \pi^0\gamma\gamma$. Again decay modes of the type $\eta \to \mu^+\mu^-$, $\eta \to \pi^0 e^+ e^-$ and $\eta \to \pi^+\pi^- e^+ e^-$ will not be considered here because they may be explained by the decays $\eta \to 2\gamma$, $\eta \to \pi^0\gamma\gamma$ and $\eta \to \pi^+\pi^-\gamma$.

For illustrations, we quote some typical Lagrangians which are required in the one-loop calculations

$$\mathcal{L}^{(\eta_8)} = i\frac{g_A M}{\sqrt{3}\,F}\,\eta_8\{(3-4a)\,(\bar{p}\gamma_5 p + \bar{n}\gamma_5 n) - (3-2a)\,(\bar{\Xi}^-\gamma_5\Xi^- + \bar{\Xi}^0\gamma_5\Xi^0)$$
$$+ 2a(\bar{\Sigma}^+\gamma_5\Sigma^+ + \bar{\Sigma}^-\gamma_5\Sigma^- + \bar{\Sigma}^0\gamma_5\Sigma^0 - \bar{\Lambda}\gamma_5\Lambda)\} \tag{4.54a}$$

$$\mathcal{L}^{(\pi^+\pi^-)} = i\,\frac{(\pi^-\partial_\mu\pi^+ - \pi^+\partial_\mu\pi^-)}{(2F)^2}\Big\{(g_A^2 - 1)\,\bar{p}\gamma^\mu p + \left[1 - g_A^2(1 - 4a(1-a))\right]\bar{\Xi}^-\gamma^\mu\Xi^-$$
$$+ 2\left[g_A^2\left(1 - 2a + \frac{4}{3}a^2\right) - 1\right](\bar{\Sigma}^+\gamma^\mu\Sigma^+ - \bar{\Sigma}^-\gamma^\mu\Sigma^-)\Big\} \tag{4.54b}$$

and

$$\mathcal{L}^{(\pi\eta_8)} = i\,\frac{(g_A a)^2}{3F^2}\,[(\eta_8\partial_\mu\pi^+ - \pi^+\partial_\mu\eta_8)\,(\bar{\Lambda}\gamma^\mu\Sigma^- - \bar{\Sigma}^+\gamma^\mu\Lambda)$$
$$+ (\eta_8\partial_\mu\pi^- - \pi^-\partial_\mu\eta_8)\,(\bar{\Lambda}\gamma^\mu\Sigma^+ - \bar{\Sigma}^-\gamma^\mu\Lambda)]. \tag{4.54c}$$

Furthermore, the interaction of the singlet meson η_0 with the baryons will be taken in the form

$$\mathcal{L}^{(\eta_0)} = i2g_0\eta_0\,\mathrm{Tr}\,(\bar{B}A_0B) = i\sqrt{\frac{2}{3}}\,g_0\eta_0$$
$$\times\{\bar{p}\gamma_5 p + \bar{n}\gamma_5 n + \bar{\Sigma}^-\gamma_5\Sigma^- + \bar{\Sigma}^+\gamma_5\Sigma^+ + \bar{\Sigma}^0\gamma_5\Sigma^0 + \bar{\Xi}^-\gamma_5\Xi^- + \bar{\Xi}^0\gamma_5\Xi^0 + \bar{\Lambda}\gamma_5\Lambda\}, \tag{4.55}$$

$$\left(A_0 = \frac{\lambda_0}{2}\gamma_5,\;\; \lambda_0 = \sqrt{\frac{2}{3}}\,1\right)$$

where g_0 is the strong coupling constant for the singlet etameson. We shall often use the constant F_0 defined by $g_0 \equiv g_A M/F_0$.

IV.3.2a. The decays $(\eta, \eta') \to 2\gamma$

Like the $\pi^0 \to 2\gamma$ decay, these decays go via anomalous triangle diagrams. Their amplitudes are given as

$$T_{\mu\nu}^{(\eta,\eta')} = -f\begin{pmatrix}\eta\\ \eta'\end{pmatrix}\varepsilon_{\mu\nu\alpha\beta}q_1{}^\alpha q_2{}^\beta, \tag{4.56}$$

where

$$f\begin{pmatrix}\eta\\ \eta'\end{pmatrix} = \frac{g_A\alpha}{\sqrt{3}\,\pi F}\begin{pmatrix}2a\cos\varphi - 2\sqrt{2}\,k\sin\varphi\\ 2a\sin\varphi + 2\sqrt{2}\,k\cos\varphi\end{pmatrix} \tag{4.57}$$

and we have set $k = F/F_0$. Table 2 collects the numerical results for the decay rates of the decays $(\eta, \eta') \to 2\gamma$ in dependence on the mixing angle φ for a value $k = -1/2$.

IV.3.2b. The decays $(\eta, \eta') \to \pi^+\pi^-\gamma$

These two processes rely on the two types of anomalous baryon loop diagrams shown in Fig. 14 [*28*]. There are box diagrams with strong interaction vertices given by the Lagrangian $\mathcal{L}_1$ (comp. eq. (2.34a), (4.54a)) and triangle diagrams with vertices given by

$\mathscr{L}_1$ and $\mathscr{L}_3$ (comp. eqs. (2.34a, c), (4.54a—c)). Triangle diagrams with a vertex $\mathscr{L}_2$ vanish by virtue of the Furry theorem.
Collecting the contributions of the box and triangle diagrams together, we obtain the amplitudes

$$T_\mu^{(\eta,\eta')} = \frac{e g_A c^{(\eta,\eta')}}{8\sqrt{3}\,F^3\pi^2}\,\varepsilon_{\mu\nu\varrho\sigma}(p^+)^\nu\,(p^-)^\varrho\,q^\sigma \tag{4.58}$$

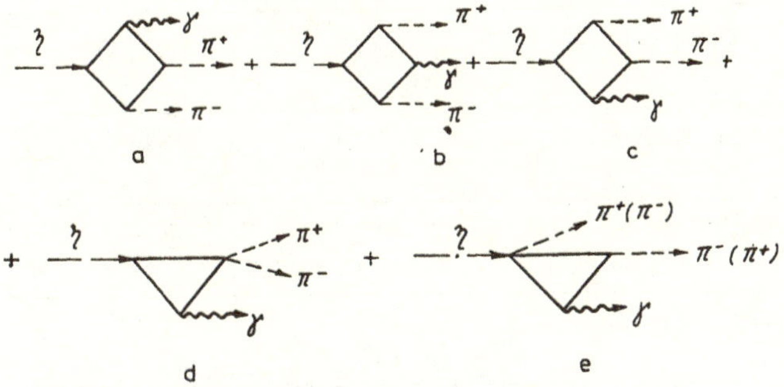

Fig. 14. Baryon loop diagrams describing the decay $\eta \to \pi^+\pi^-\gamma$

where p^+, p^- and q are the momenta of the π^+, π^- mesons and the photon, respectively. The contribution of different members of the baryon octet to the loops are accounted for in factors $c^{\eta,\eta'}$,

$$\begin{aligned} c^\eta &= c_8 \cos\varphi - c_0 k \sin\varphi \\ c^{\eta'} &= c_8 \sin\varphi + c_0 k \cos\varphi, \end{aligned} \tag{4.59}$$

where

$$\begin{aligned} c_8 &= 6a - g_A{}^2(1 + (2a-1)^3) \\ c_0 &= \sqrt{2}\left[3 - g_A{}^2\left(1 - 2a + \frac{14}{9}a^2\right)\right]. \end{aligned} \tag{4.60}$$

Eq. (4.58) yields the decay width

$$\Gamma_{\binom{\eta}{\eta'}\to\pi^+\pi^-\gamma} = \frac{\alpha}{2}\left(\frac{g_A c^{(\eta,\eta')}}{6}\right)^2\left(\frac{m_{\eta,\eta'}}{4\pi F}\right)^6\left[1 - \left(\frac{2m_\pi}{m_{\eta,\eta'}}\right)^2\right]^5 m_{\eta,\eta'} I_{\eta,\eta'}. \tag{4.61}$$

Here $I_\eta = 0.075$ and $I_{\eta'} = 0.46$ are the values of the relevant phase space integrals. Note that in order to account for the ϱ-resonance in the $\pi^+\pi^-$ final state, we have modified our matrix elements by a form factor taken from the work of GOUNARIS and SAKURAI [69]. The predictions resulting from eq. (4.61) are quoted in Table 2 for different values of the mixing angles $\varphi = -11°$ and $\varphi = -18°$.
For comparison, we indicate also the analogous formulas for the decay amplitudes calculated within the quark model [70, 71]. In the case of the $(\eta, \eta') \to 2\gamma$ decay we have[17])

$$f\begin{pmatrix}\eta\\ \eta'\end{pmatrix} = \frac{\alpha}{\sqrt{3}\,\pi F}\begin{pmatrix}\cos\varphi - 2\sqrt{2}\,\bar{k}\sin\varphi \\ \sin\varphi + 2\sqrt{2}\,\bar{k}\cos\varphi\end{pmatrix}, \qquad \bar{k} = F/\bar{F}_0 \tag{4.62}$$

[17]) Using the Lagrangian (2.37), the best estimates of the radiative neutral meson decays considered here have been found for $g_A' = 1$ and $\bar{F}_0 = F$, i.e. $\bar{k} = 1$ [71]. We shall use these values henceforth. For $g_A' = 1$ only the usual Yukawa vertex $\mathscr{L}_1$ contributes to the processes considered above.

Table 2

Decay widths for the radiative decays of the neutral pseudoscalar mesons (π^0, η, η') [*71*]. The width for the decay $\eta \to \pi^+\pi^-\gamma$ has been calculated by using two different resonance factors taken from ref. [*70*] (values in brackets) and ref. [*69*]

		$\eta \to 2\gamma$	$\eta \to \pi^+\pi^-\gamma$	$\eta' \to 2\gamma$	$\dfrac{\Gamma(\eta' \to \rho^0\gamma)}{\Gamma(\eta' \to 2\gamma)}$
Meson-baryon model ($k = -0.5$)	$\varphi = -18°$	200 eV	40 eV (34 eV)	4.2 keV	~ 16
	$\varphi = -11°$	280 eV	55 eV (45 eV)	3.7 keV	~ 15
Quark loop model ($\bar{k} = 1$)	$\varphi = -18°$	550 eV	70 eV (55 eV)	5 keV	~ 6.6
	$\varphi = -11°$	380 eV	55 eV (44 eV)	5.9 keV	~ 7.5
Experiment		(323 ± 54) eV [*60*]	(41.6 ± 7) eV [*60*]	(5.9 ± 1.6) keV [*65a*] (5.4 ± 2.1) keV [*65b*]	~ 15

while for the $(\eta, \eta') \to \pi^+\pi^-\gamma$ decays the coefficients $c^{(\eta,\eta')}$ take the form

$$
\begin{aligned}
c^{\eta} &= 2\cos\varphi - 2\sqrt{2}\,\bar{k}\sin\varphi \\
c^{\eta'} &= 2\sin\varphi + 2\sqrt{2}\,\bar{k}\cos\varphi\,.
\end{aligned}
\tag{4.63}
$$

At this point a remark on the constant F_0 is in order, the value of which turns out to be model-dependent. This fact should not cause any confusion if one recognizes that, in contrast to the model-independent octet decay constant F describing weak leptonic decays $\pi \to \mu\nu_\mu$, $\pi^0 \to e^+e^-$, $\eta_8 \to \mu^+\mu^-$ etc.[18]) F_0 (or $\bar{F}_0$) plays the role of an auxiliary parameter which is associated to the strong interaction constant $g_0(\bar{g}_0)$. Indeed, since there exist no neutral weak decays of the type $\eta_0 \to \mu^+\mu^-$, $\nu\bar{\nu}$ etc. measuring F_0 (or $\bar{F}_0$), this parameter has no independent physical meaning but rather depends on the strong interaction model chosen.

The predictions for the rates of the decays $(\eta, \eta') \to 2\gamma$ and $(\eta, \eta') \to \pi^+\pi^-\gamma$ are given in Table 2 for different values $\varphi = -11°$ or $18°$. Obviously, at $\varphi = -11°$ the quark model results for eta-decays are in good agreement with experimental data. If one accepts on the other hand, the new value of the mixing angle, $\varphi = -18°$, the quark model predicts an increase of about 60% in the η-decay rates as compared to the present data [*60*] (see also [*72*]).

IV.3.2c. The decay $\eta \to \pi^0\gamma\gamma$ [*73*]

Let us conclude the investigation of the eta-decays with the process $\eta \to \pi^0\gamma\gamma$. This decay differs strongly from other meson decays as it proceeds only via meson loops. Thus, the situation is similar to that encountered in the case of the π^0 polarizability.

---

[18]) The neutral weak decays are, of course, superimposed from the more significant electromagnetic contributions.

We remind the reader that in the one-loop approximation the dominant contributions to meson decay processes come, in general, from the baryon loop diagrams. Meson loop diagrams either do not contribute (if we have an odd number of external meson lines in a given diagram) or their contribution is small when compared with baryon loops so that it can be neglected. As we found, an important exception of this rule happens, however, for the above decay where all baryon diagrams turn out to cancel each other due to the chiral symmetry. Thus, the only nonvanishing contributions come from the meson loops. Consequently, the decay probability for the process $\eta \to \pi^0\gamma\gamma$ should be small when compared with the decays $\eta \to 2\gamma$ or $\eta \to \pi^+\pi^-\gamma$ going via baryon loops. The present

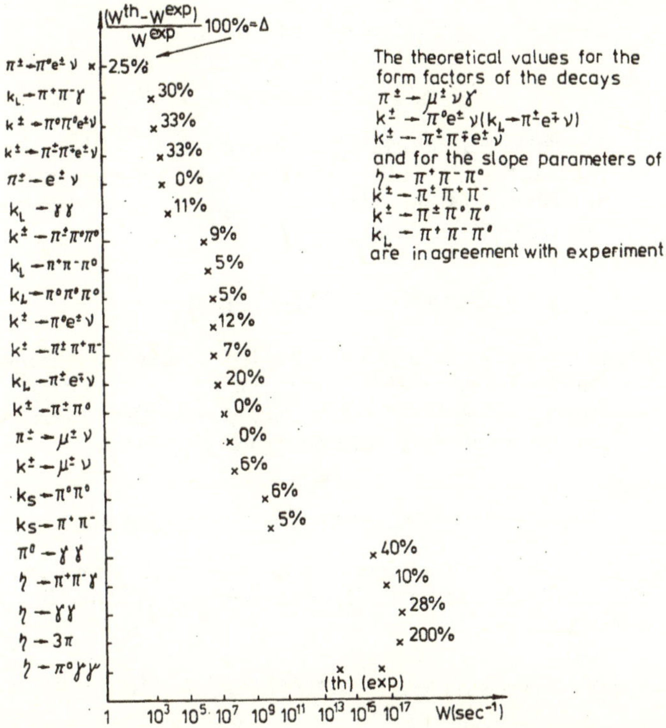

Fig. 15. Decay probabilities of the main decay modes of the meson octet [28]. The figure shows the relative errors of the theoretical predictions obtained on the basis of the chiral theory. The last process $\eta \to 3\pi$ is directly related to the symmetry breaking mechanism and thus leads to an exceptionally large error

experimental data for the decays $\eta \to \pi^0\gamma\gamma$ and $\eta \to \pi^+\pi^-\gamma$ do, however, not confirm this conjecture.

Since the above decay represents the only case among a large number of decays of the meson octet (comp. Fig. 15) where a sharp disagreement of the predictions of the chiral theory with data is found, it seems to be useful to check once more the experimental situation. If one nevertheless wants to believe in the data, there arises the necessity to find other kinds of dominating decay mechanisms to describe this decay.

Let us briefly consider the contributions of the meson loops. The $\eta-\pi$ interaction comes from the symmetry-breaking term in the meson Lagrangian (cf. eq. (3.21)). We have

$$\Delta\mathcal{L} = \frac{m_\pi^2}{\sqrt{3}} :\pi^0\eta_8\left(1 - \frac{\pi^2}{6F^2}\right): + \frac{m_\pi^2}{12F^2} :\pi^2\eta_8^2: + \frac{m_\pi^2}{24F^2}:(\pi^2)^2:. \tag{4.64}$$

The total expression arising from the meson loop diagrams reads

$$T_{\eta\to\pi^0\gamma\gamma} = -i\,\frac{2}{\sqrt{3}}\left(\frac{\pi e}{F}\right)^2 \varepsilon_1{}^\mu\varepsilon_2{}^\nu(g_{\mu\nu}q_1q_2 - q_{1\nu}q_{2\mu})\,\frac{\left(q_1q_2 - \frac{2}{3}\,m_\pi{}^2\right)}{m_\eta{}^2 - m_\pi{}^2}\,I\left(\frac{q_1q_2}{2m_\pi{}^2}\right) \tag{4.65}$$

where q_1, q_2, $\varepsilon_1{}^\mu$ and $\varepsilon_2{}^\nu$ are the momenta and polarization fourvectors of the two emitted photons, and the function $I(\xi)$ is defined in eq. (4.23).

The total decay width of the decay $\eta \to \pi^0\gamma\gamma$ evaluated with eq. (4.65) is

$$\Gamma_{(\eta\to\pi^0\gamma\gamma)} = 10^{-2}\ \text{eV} \tag{4.66}$$

which differs by three orders of magnitude from the experimental data quoted in the Rosenfeld table [60].

We have thus seen that within the chiral model, the decay $\eta \to \pi^0\gamma\gamma$ cannot be reduced to the decay $\eta \to 2\gamma$ by inserting an additional strong (pion) vertex into the corresponding loop diagram. The decay is rather related to the decay $\eta \to \pi^0\pi^+\pi^-$ where charged pions subsequently annihilate into a photon pair.

IV.3.3. Kaon decays [28]

In this subsection we discuss the most characteristic decays of charged and neutral kaons. In contrast with the decays of the eta-mesons, all the kaon decays are obviously connected with weak interactions. On the other hand, as we have seen, the pions occupied some intermediate place decaying both electromagnetically ($\pi^0 \to 2\gamma$) and weakly.

IV.3.3a. Leptonic decays: $K^\pm \to \mu^\pm\nu_\mu(e^\pm\nu_e)$

Let us begin the investigation of the decays with the most characteristic decay $K^\pm \to \mu^\pm\nu_\mu(e^\pm\nu_e)$. Using the weak interaction Lagrangian (cf. eqs. 3.54, 55, 59) one obtains in the tree approximation

$$T_{(K^+\to\mu^+\nu_\mu)} = i\,\frac{G}{\sqrt{2}}\sin\Theta\left(\sqrt{2}\,Fk^\mu l_\mu{}^{(+)}\right), \tag{4.67}$$

where k_μ is the kaon momentum, $l_\mu{}^{(+)}$ the leptonic current (comp. eq. (4.40)) and Θ the Cabibbo angle. The amplitude (4.67) leads to the decay probability

$$W_{(K^+\to\mu^+\nu_\mu)} = \sin^2\Theta\,\frac{(GF)^2}{4\pi}\,m_K m_\mu{}^2\left(1 - \frac{m_\mu{}^2}{m_K{}^2}\right)^2 = 4.8\cdot 10^7\ \text{s}^{-1} \tag{4.68}$$

whereas the experimental value is [60]

$$W^{\text{exp}}_{(K^+\to\mu^+\nu_\mu)} = 5.1\cdot 10^7\ \text{s}^{-1}. \tag{4.69}$$

Eqs. (4.41) and (4.68) show that for $\Theta \approx 0.26$ the constants F_π and F_K can indeed be chosen equal.

As for the analogous decay $\pi^+ \to \mu^+\nu_\mu$, one can take into account one-loop corrections arriving at an expression like (4.47) with m_π replaced by m_K. Although the loop corrections are now larger than in the pion case, they lie within our limit of accuracy of 20% to 30% and will therefore be neglected.

IV.3.3b. Semileptonic decays: K_{l3}, K_{l4} decays.

K_{l3} decay

In this subsection we shall discuss the semileptonic decays $K^- \to \pi^0 l\bar{\nu}$, $K_L \to \pi^+ l\bar{\nu}$ and $K^- \to \pi^+\pi^- l\bar{\nu}$, $K_L \to \pi^0\pi^+ l\bar{\nu}$ usually called K_{l3} and K_{l4} decays.
The above two K_{l3} decays lead in the tree approximation to the decay amplitude

$$T^{(0)}_{m_i} = \frac{G}{\sqrt{2}} a_i (p_{m_i} + p_\pi)^\mu l_\mu^{(+)} \tag{4.70}$$

where p_{m_i} is the momentum of the decaying particle and

$$a_1 = \frac{\sin \Theta}{\sqrt{2}}, \qquad a_2 = \sin \Theta .$$

The corresponding decay probabilities are

$$W_{K_{Ll3}} = 9 \cdot 10^6 \text{ s}^{-1}, \qquad W_{K^\pm l3} = 4.2 \cdot 10^6 \text{ s}^{-1} \tag{4.71}$$

while the experimental values are [60]

$$W^{\text{exp}}_{K_L l3} = 7.5 \cdot 10^6 \text{ s}^{-1}, \qquad W^{\text{exp}}_{K^\pm l3} = 3.9 \cdot 10^6 \text{ s}^{-1}. \tag{4.72}$$

Let us write down the most general expression for the K_{l3} amplitude. We have

$$T_{K^- \to \pi^0 l\bar{\nu}} = i \frac{G \sin \Theta}{2} \{(p_K + p_\pi)^\mu f_+(q^2) + (p_K - p_\pi)^\mu f_-(q^2)\} \bar{l}_\mu \tag{4.73}$$

where $q^2 = (p_K - p_\pi)^2$. $f_\pm(q^2)$ are form factors usually parametrized in the form

$$f_\pm(q^2) = f_\pm(0) \left\{1 + \lambda_\pm \frac{q^2}{m_\pi^2}\right\}, \tag{4.74}$$

and $\lambda_\pm$ are so-called slope parameters. To study the effect of $SU(3)$ breaking it is also convenient to consider the following combination of f_+ and f_-

$$f_0 = f_+ + \frac{q^2}{m_K^2 - m_\pi^2} f_- = f_0(0) \left\{1 + \lambda_0 \frac{q^2}{m_\pi^2}\right\}.$$

In the case of exact $SU(3)$ symmetry the one-loop approximation contributes only to the form factors $f_\pm(q^2)$, but not to $f_0(q^2)$. Thus, to get also a nontrivial result for $f_0(\lambda_0 \neq 0)$, one must include $SU(3)$ symmetry breaking effects. Let us consider the simplest possibility of choosing physical baryon masses leaving the relations between coupling constants unchanged. In the one-loop approximation one gets [75] (comp. Fig. 16a–c)

$$f_+(q^2) = 0.95 + 0.024 \frac{q^2}{m_\pi^2}; \qquad f_-(q^2) = -0.63 + 0.004 \frac{q^2}{m_\pi^2} \tag{4.75a}$$

and thus

$$f_0(q^2) = 0.95 - 0.027 \frac{q^2}{m_\pi^2}. \tag{4.75b}$$

From these formulas we get the slope parameters

$$\lambda_+ \approx 0.025, \qquad \lambda_0 \approx -0.028$$

which are in good agreement with the experimental values [76]

$$\lambda_+^{\text{exp}} = 0.025 \pm 0.007; \qquad \lambda_0^{\text{exp}} = -0.017 \pm 0.011.$$

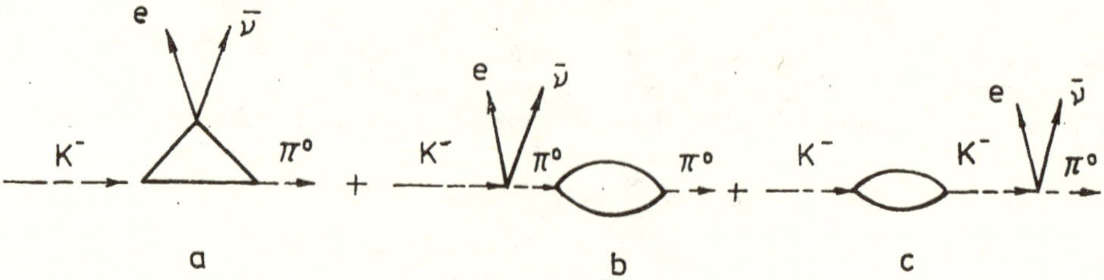

Fig. 16. Baryon loop diagrams contributing to the form factor of the K_{l3} decay

K_{l4} decay [77]

The K_{l4}-decay $K^+ \to \pi^+\pi^-e^+\nu_e$ is interesting by itself and due to the possibility to extract from it information on many other important physical quantities (e.g. the scattering lengths of the $\pi\pi$-system). This is also the reason why this process has been discussed so often in the literature [1]. The decay amplitude contains four form factors three of them are axial ones and one is a vector form factor. The axial form factors may be estimated, for instance, in the framework of the standard current algebra approach [1] or by making use of the tree approximation of the chiral theory [2]. On the other hand, the description of the vector form factor is much more involved. The standard soft-pion techniques of current algebra are now useless because the vector part of the amplitude vanishes for vanishing pion momenta. Also the tree approximation of the chiral theory does not provide any information on the vector form factor. By these reasons one expects that now the one-loop approximation should play an important role.

The amplitude of the K_{l4} decay is usually written in the form

$$T_{K_{l4}} = \sin\Theta \frac{G}{\sqrt{2}\, m_K} \Big\{ f(p^+ + p^-)^\mu + g(p^+ - p^-)^\mu + r(p^K - p^+ - p^-)^\mu + i \frac{h}{m_K^2} \varepsilon^{\mu\nu\varrho\sigma} p_\nu^K (p^+ + p^-)_\varrho (p^+ - p^-)_\sigma \Big\} l_\mu^{(-)} \qquad (4.76)$$

where p^+, p^- and p^K are the momenta of the π^+, π^- and K^+ meson, respectively, f, g and r are the axial form factors and h is the vector form factor.

The axial form factors get their main contributions from the tree approximation (see Fig. 17a, b). Diagram 17a may be calculated by using the Lagrangian

$$\mathscr{L}^{(1)}_{\text{weak}} = \sin\Theta \frac{G}{3F} [\pi^-(K^+\partial_\mu\pi^+ + \pi^+\partial_\mu K^+) - 2K^+\pi^+\partial_\mu\pi^-]\, l^{\mu(-)}. \qquad (4.77)$$

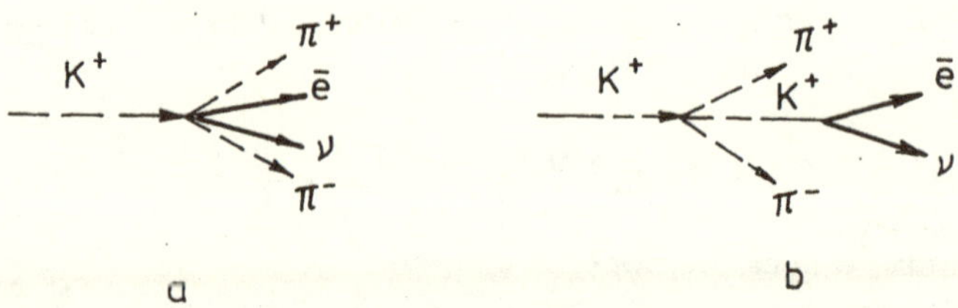

Fig. 17. Tree diagrams contributing to the axial form factors of K_{l4} decay

This diagram determines almost completely the value of the decay probability and the form factors f and g. It gives rise to the amplitude

$$T^{(1)}_{K_{l4}} = \sin \Theta \frac{G}{3F} (3p^+ + q)^\mu l_\mu^{(-)} \tag{4.78}$$

where $q = p^K - p^+ - p^-$. To get also a reasonable expression for the third form factor r, implies considering the diagram 17b[19]). It can be evaluated by using the expressions for the weak decay vertex and the corresponding $\pi\pi$KK vertices which follow from the general expressions $\mathcal{L}^{(1)}_{\text{weak}}$, $\mathcal{L}_{\text{inv}}$ and $\Delta\mathcal{L}_m$ (cf. (3.54), (2.32), (3.21)). This diagram contributes as follows

$$T^{(2)}_{K_{l4}} = \sin \Theta \frac{G}{6F} \left(1 + 6 \frac{p^K p^-}{q^2 - m_K{}^2}\right) q^\mu l_\mu^{(-)} . \tag{4.79}$$

Hence, we derive the following expressions for the axial form factors

$$f = g = \frac{m_K}{\sqrt{2}\, F}; \qquad r = \sqrt{2}\, \frac{m_K}{F} \frac{q p^+}{m_K{}^2 - q^2} = \begin{cases} 0, & p^+ = 0 \\ \dfrac{m_K}{\sqrt{2}\, F}, & p^- = 0 \end{cases} \tag{4.80}$$

which are in complete agreement with the predictions from current algebra [1].
The evaluation of the vector form factor h completely parallels the evaluation of the decay amplitudes of the processes $\eta \to \pi^+\pi^-\gamma$ and $K_L \to \pi^+\pi^-\gamma$ (cf. Sect. IV.3.2b and IV.3.3c). h receives contributions from box diagrams with strong vertices given by $\mathcal{L}_1$ and triangle diagrams with one strong vertex $\mathcal{L}_1$ and one strong vertex $\mathcal{L}_2$ (comp. Figs. 18a—c). Fig. 18a represents schematically three types of box diagrams where the lepton

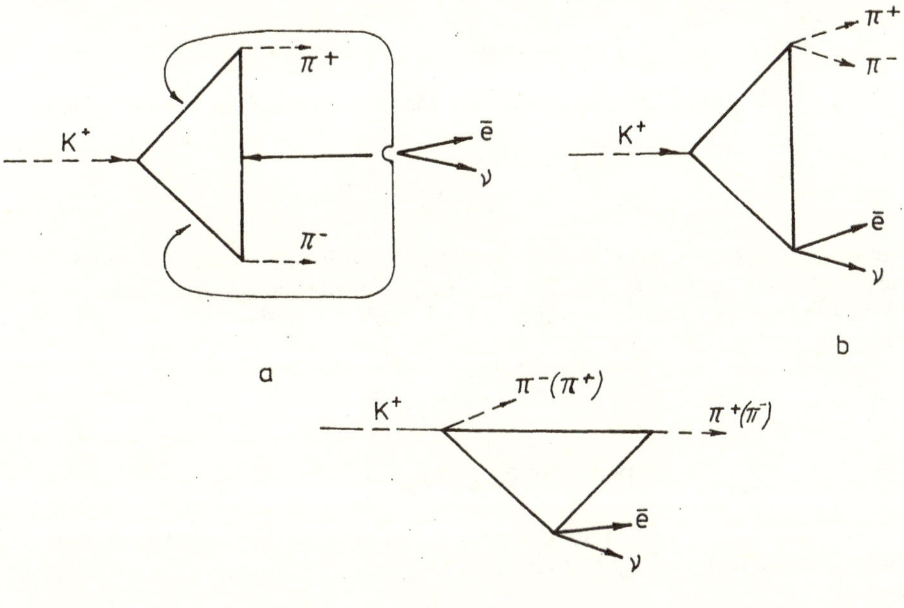

Fig. 18. Baryon loop diagrams contributing to the vector form factor of K_{l4} decay (Fig. 18a is similar to Figs. 14a, b, c)

[19]) The importance of this diagram for the evaluation of r has been first recognized by WEINBERG [1].

pair is emitted either from a baryon propagator between the K^+ and π^+ mesons, the K^+ and π^- mesons or between the π^+ and π^- mesons.
Retaining in the loop integral of diagram 18a only those terms which do not depend on external momenta, we arrive at the expression

$$T^{\Box} = i \sin \Theta \frac{g_A G c^{\Box}}{(4\pi)^2 F^3} \varepsilon^{\mu\nu\varrho\sigma} p_\nu{}^K p_\varrho{}^- p_\sigma{}^+ l_\mu{}^{(-)} \tag{4.81}$$

where

$$c^{\Box} = -24 g_A{}^2 a \left(1 - 2a + \frac{4}{3} a^2\right) \approx -6.5.$$

On the other hand, the baryon triangle diagrams (Fig. 18b—c) lead to the following expressions for the coefficients c

$$\begin{aligned} c^{\Delta}_{b(\pi^+\pi^-)} &= c^{\Delta}_{c(K^+\pi^-)} = 6a \left[-1 + g_A{}^2 \left(3 - 6a + \frac{32}{9} a^2\right)\right], \\ c^{\Delta}_{c(K^+\pi^+)} &= \frac{16}{3} g_A{}^2 a^3, \qquad c^{\Delta}_{b+c} \approx 1.7. \end{aligned} \tag{4.82}$$

Summing up all baryon loop diagrams, the total contribution to the vector part of the K_{l4}-decay amplitude reads

$$T^{(h)}_{K_{l4}} = \sin \Theta \frac{g_A G c}{(4\pi)^2 F^3} \varepsilon^{\mu\nu\varrho\sigma} p_\nu{}^K (p^+ + p^-)_\varrho (p^+ - p^-)_\sigma l_\mu{}^{(-)} \tag{4.83}$$

with[20])

$$c = 6a - g_A{}^2[1 + (2a - 1)^3] \approx 2.4.$$

Eq. (4.83) provides the following expression for the vector form factor h (comp. eq. (4.76)) [77]

$$h = -\frac{g_A c}{\sqrt{2}\, \pi^2} \left(\frac{m_K}{2F}\right)^3. \tag{4.84}$$

In order to compare the above results with the experimental data, let us give first a rough estimate of the probability of the K_{l4} decay. To this end, consider only those parts of the amplitude (4.76) which contain the form factors f and g and use eq. (4.80). We then find

$$W^{(\text{tree})}_{K_{l4}} \simeq 1.7 \cdot 10^3 \text{ s}^{-1} \tag{4.85}$$

while the experimental value is

$$W^{\text{exp}}_{K_{l4}} = (3.26 \pm 0.15) \cdot 10^3 \text{ s}^{-1}.$$

Keeping in mind that the tree approximation is rather crude, the theoretical value (4.85) may be considered to be of right order[21]).

---

[20]) Note that the coefficient c is equal to the value of an analogous coefficient arising in the decay $\eta \to \pi^+\pi^-\gamma$ (comp. eq. (4.60)).

[21]) Preliminary calculations taking into account convergent box diagrams only (in general the main contribution) show an increase in f and g of about 30% which is in the right direction when compared with the data.

Let us now compare the values of h and of the ratio $h/f^{\exp}$ which is often used in experimental papers, with the data. Taking $f^{\exp} = (6.15 \pm 0.15)$ [78a], we find

$$h = -3.7, \qquad h/f^{\exp} = -0.6 \tag{4.86}$$

whereas the data give

a) $h^{\exp} = -2.95 \pm 0.75$; $(h/f)^{\exp} = -0.48 \pm 0.12$ [78a]
b) $(h/f)^{\exp} = -0.71 \pm 0.23$ [78b]
c) $(h/f)^{\exp} = -0.97 \pm 0.46$ [78c].

For completeness, let us also quote the theoretical estimates for h obtained in a model with ϱ dominance [79]

$$1.5 \leqq |h| \lesssim 5$$

and the value $h = -3.7$ (see [80a]) in a current algebra model (see also [80b]). Finally, evaluating the K_{l4} decay via anomalous quark loops yields instead of eq. (4.84) [71]

$$h = -\frac{\sqrt{2}}{\pi^2}\left(\frac{m_K}{2F}\right)^3 = -2.5 \tag{4.87}$$

which is also in reasonable agreement with the experimental data.

IV.3.3c. Radiative decays: $K_L \to 2\gamma$, $K_L \to \pi^+\pi^-\gamma$ [28, 81].

The decay $K_L \to 2\gamma$

The radiative decays of the K_L meson may be directly related to the radiative decays of the π^0- and η-mesons by considering the pole diagrams of Fig. 19 [82]. To describe the direct weak transitions $K_L \to \pi^0$, η_8 and η_0, we use the following Lagrangian proposed by Sakurai [83]

$$\mathcal{L}^{(K_L)} = \sqrt{2}\, GF^2\, \partial^\mu K_L\, \partial_\mu\left(\pi^0 + \frac{\eta_8}{\sqrt{3}} + \sqrt{\frac{2}{3}}\,\eta_0\right) \simeq A K_L\left(\pi^0 + \frac{\eta_8}{\sqrt{3}} + \sqrt{\frac{2}{3}}\,\eta_0\right) \tag{4.88}$$

$$A = \sqrt{2}\, GF^2 m_K{}^2 = 3.6 \cdot 10^{-2}\ (\mathrm{MeV})^2.$$

The decay probability of the $K_L \to 2\gamma$ decay may then be expressed by straightforward calculations in the form

$$W_{(K_L\to 2\gamma)} = \frac{3A^2}{m_K{}^4}\left\{\frac{1}{1-\dfrac{m_\pi{}^2}{m_K{}^2}} + \frac{1}{3\left(1-\dfrac{m^2_{\eta_8}}{m_K{}^2}\right)} + \frac{2k}{3a\left(1-\dfrac{m^2_{\eta_0}}{m_K{}^2}\right)} + \frac{\sin 2\varphi\, m_K{}^2(m^2_{\eta'} - m_\eta{}^2)\,(a+k)}{3\sqrt{2}\,a(m^2_{\eta_8} - m_K{}^2)\,(m^2_{\eta_0} - m_K{}^2)}\right\}^2 W^{(m_{\eta_8}\to m_K)}_{(\eta_8\to 2\gamma)}. \tag{4.89}$$

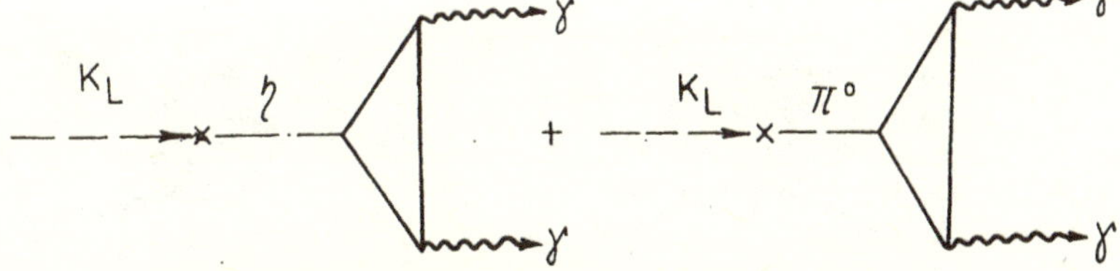

Fig. 19. Baryon loop diagrams contributing to the decay $K_L \to 2\gamma$

Here $W^{(m_{\eta_8} \to m_K)}_{(\eta_8 \to 2\gamma)}$ is the probability of the decay $\eta_8 \to 2\gamma$ where the mass of the η_8-meson is replaced by the mass of the K_L meson. Using the mass relations

$$m_{\eta_0}^2 = m_{\eta'}^2 \cos^2 \varphi + m_\eta{}^2 \sin^2 \varphi ,$$

$$m_{\eta_8}^2 = m_{\eta'}^2 \sin^2 \varphi + m_\eta{}^2 \cos^2 \varphi$$

and taking into account the expression

$$W^{(m_{\eta_8} \to m_{K_L})}_{(\eta_8 \to 2\gamma)} = \frac{m_K}{3\pi} \left(\frac{\alpha a g_A m_K}{4\pi F} \right)^2 , \tag{4.90}$$

we finally get ($k = -0.5$)

$$W_{(K_L \to 2\gamma)} = 8 \cdot 10^3 \text{ s}^{-1} . \tag{4.91}$$

The experimental value is [*60*]

$$W^{\text{exp}}_{(K_L \to 2\gamma)} = (9.45 \pm 0.96) \cdot 10^3 \text{ s}^{-1} . \tag{4.92}$$

In the quark model the analogous calculations give [*71*] $W_{(K_L \to 2\gamma)} = 14 \cdot 10^3 \text{ s}^{-1}$. The above results refer to the mixing angle $\varphi = -18°$.

The decay $K_L \to \pi^+\pi^-\gamma$

The decay probability of this decay takes the form (comp. Fig. 20)

$$W_{(K_L \to \pi^+\pi^-\gamma)} = \frac{A^2}{6 m_K{}^4} \left\{ \frac{1}{1 - \frac{m_\pi{}^2}{m_K{}^2}} + \frac{1}{3\left(1 - \frac{m_{\eta_8}^2}{m_K{}^2}\right)} + \frac{\sqrt{2}\, k c_0}{3 c_8 \left(1 - \frac{m_{\eta_0}^2}{m_K{}^2}\right)} \right.$$
$$\left. + \frac{\sin 2\varphi\, m_K{}^2 (m_{\eta'}^2 - m_\eta{}^2) (\sqrt{2} + k c_0/c_8)}{6 (m_{\eta_8}^2 - m_K{}^2)(m_{\eta_0}^2 - m_K{}^2)} \right\} \alpha \left(\frac{g_A c_8}{2} \right)^2 \left(\frac{m_K}{4\pi F} \right)^6 \left[1 - \left(\frac{2 m_\pi}{m_K} \right)^2 \right]^5$$
$$\times m_K I_{m_K} \tag{4.93}$$

where $I_{m_K} = 0.063$, and the coefficients c_8, c_0 are given in eqs. (4.60). For $\varphi = -18°$ we obtain

$$W_{(K_L \to \pi^+\pi^-\gamma)} \approx 300 \text{ s}^{-1} . \tag{4.94}$$

The same value follows also in the quark model. The experimental value is a bit larger [*84*]

$$W^{\text{exp}}_{(K_L \to \pi^+\pi^-\gamma)} = (558 \pm 54) \text{ s}^{-1} . \tag{4.95}$$

Note that this enlarged value can also be derived in the quark model if we choose $\varphi = -11°$.

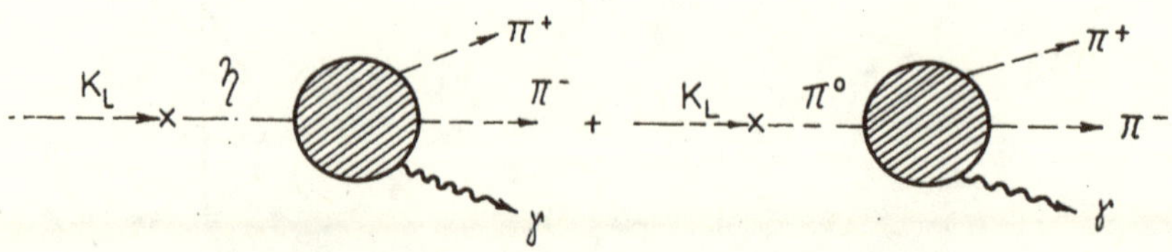

Fig. 20. Baryon loop diagrams describing the decay $K_L \to \pi^+\pi^-\gamma$

IV.3.4. Weak decays of charmed mesons [21, 24]

In this subsection we want to utilize the meson part of the $SU(4) \times SU(4)$ Lagrangian given in Sect. II.3 in order to get a rough estimate for some typical leptonic and semi-leptonic decays of F and D mesons. Throughout we confine ourselves to the tree approximation which is sufficient for this purpose. If more detailed experimental information on the structure constants of charmed meson decays (e.g. form factors of D_{l3}, D_{l4} decays etc.) is available, necessary one-loop calculations can easily be performed. To minimize troubles due to possible $SU(4)$ violating in the meson-baryon coupling constants, such calculations should be performed most profitably within the $SU(4) \times SU(4)$ meson-quark Lagrangian.

IV.3.4a. Leptonic decays: $F \to \mu\bar{\nu}_\mu$, $D \to \mu\bar{\nu}_\mu$

Let us first consider the leptonic decays $F^- \to \mu\bar{\nu}_\nu$, $D^- \to \mu\bar{\nu}_\mu$. The relevant part of the effective Lagrangian (3.50) now reads

$$\mathcal{L}^{(1)}_{\text{weak}} = GF(\cos\Theta\, \partial_\mu F^- - \sin\Theta\, \partial_\mu D^-)\, J_l^{\mu,W}. \tag{4.96}$$

The decay amplitudes and probabilities take a form analogous to eqs. (4.67, 4.68). Using $m_D \approx 1.87$ GeV, $m_F \approx 2.03$ GeV, we obtain

$$W_{(D^-\to\mu\bar{\nu}_\mu)} = 2\cdot 10^8\ \text{s}^{-1}, \qquad W_{(F^-\to\mu\bar{\nu}_\mu)} = 3\cdot 10^9\ \text{s}^{-1}. \tag{4.97}$$

IV. 3.4b. Semileptonic decays: D_{l3}, F_{l3} and D_{l4}, F_{l4} decays

D_{l3}, F_{l3} *decays*

The D_{l3} and F_{l3} decays are described by the following part of the effective weak Lagrangian

$$\mathcal{L}^{(2)}_{\text{weak}} = -i\frac{G}{\sqrt{2}}\left\{\cos\Theta\left(\bar{D}^0 \overleftrightarrow{\partial}_\mu K^- + \sqrt{\frac{2}{3}}\,\eta_8 \overleftrightarrow{\partial}_\mu F^-\right)\right.$$
$$\left. + \sin\Theta\left(-\bar{D}^0\overleftrightarrow{\partial}_\mu \pi^- + \frac{1}{\sqrt{6}}\,\eta_8\overleftrightarrow{\partial}_\mu D^- + K^0\overleftrightarrow{\partial}_\mu F^-\right)\right\} J_l^{\mu,W}. \tag{4.98}$$

The amplitudes of the Cabibbo favoured reactions $\bar{D}^0 \to K^+ e\bar{\nu}_e$ and $F^- \to \eta e\bar{\nu}_e$ follow from eq. (4.98) to be[22])

$$T_{(\bar{D}^0\to K^+e\bar{\nu}_e)} = -\frac{G}{\sqrt{2}}\cos\Theta(p_D + p_K)^\mu\, l_\mu^{(+)} \tag{4.99}$$

$$T_{(F^-\to\eta e\bar{\nu}_e)} = \frac{G}{\sqrt{3}}\cos\Theta\,(p_F + p_\eta)^\mu\, l_\mu^{(+)}. \tag{4.100}$$

Then

$$W_{(\bar{D}^0\to K^+e\bar{\nu}_e)} = \frac{G^2 m_D m_K{}^4 \cos^2\Theta}{12\pi^3}\, I(\varkappa_{D,K}) \tag{4.101}$$

where $I(\varkappa_{D,K})$ is the phase space integral

$$I(\varkappa_{D,K}) = \int_1^{\varkappa_{D,K}} dx(x^2 - 1)^{3/2}$$

---

22) In the following estimates we neglect $\eta_0 - \eta_8$ mixing.

and $\varkappa_{D,K} = (m_D^2 + m_K^2)\, 2m_D m_K$. Eq. (4.101) leads to the prediction

$$W_{(\bar{D}^0 \to K^+ e\bar{\nu}_e)} = 10^{11}\ \mathrm{s}^{-1}. \tag{4.102}$$

The same estimate holds for the decay $D^- \to K^0 e\bar{\nu}_e$. Similarly, one obtains[23])

$$W_{(F^- \to \eta e\bar{\nu}_e)} = 10^{11}\ \mathrm{s}^{-1}.$$

Finally, we estimate the Cabibbo forbidden decays $\bar{D}^0 \to \pi^+ e\bar{\nu}_e$, $D^- \to \eta e\bar{\nu}_e$ and $F^- \to \bar{K}^0 e\bar{\nu}_e$ that are suppressed by a factor $\tan^2 \Theta \approx 0.05$ when compared to the corresponding favoured reactions. We have

$$\begin{aligned} W_{(D^{-0} \to \pi^+ e\bar{\nu}_e)} &\simeq 1.2 \cdot 10^{10}\ \mathrm{s}^{-1} \\ W_{(D^- \to \eta e\bar{\nu}_e)} &\simeq 1.1 \cdot 10^{9}\ \mathrm{s}^{-1} \\ W_{(F^- \to \bar{K}^0 e\bar{\nu}_e)} &\simeq 1.2 \cdot 10^{10}\ \mathrm{s}^{-1}. \end{aligned} \tag{4.103}$$

The above predictions are in agreement with calculations performed in the quark model [*85*]. Analogous results have also been derived within a chiral $SU(4) \times SU(4)$ meson theory using an empirical current $\times$ current Lagrangian [*35*].

D_{l4}, F_{l4} *decays*

The D_{l4} and F_{l4} decays $\bar{D}^0 \to K^0 \pi^+ e\bar{\nu}_e$, $F^- \to K^+ K^- e\bar{\nu}_e$ can be calculated in the tree approximation from the effective Lagrangian

$$\begin{aligned} \mathcal{L}^{(3)}_{\text{weak}} = \cos\Theta \frac{G}{3F} [&(2\bar{D}^0 \bar{K}^0\, \partial_\mu \pi^- - \pi^- \bar{D}^0\, \partial_\mu \bar{K}^0 - \pi^- \bar{K}^0\, \partial_\mu \bar{D}^0) \\ &+ (2F^- K^-\, \partial_\mu K^+ - F^- K^+\, \partial_\mu K^- - K^- K^+\, \partial_\mu F^-)]\, J_l^{\mu,W}. \end{aligned} \tag{4.104}$$

This yields

$$T_{(\bar{D}^0 \to K^0 \pi^+ e\bar{\nu}_e)} = -i \cos\Theta \frac{G}{3F} p^\mu l_\mu^{(+)}$$

where

$$p = p_{\bar{D}^0} + 2p_{\pi^+} - p_{K^0} = 3p_{\pi^+} + p_l, \tag{4.105}$$

p_l being the momentum of the lepton pair. From eq. (4.105) we obtain the decay probability

$$W_{(\bar{D}^0 \to K^0 \pi^+ e\bar{\nu}_e)} = \frac{\cos^2 \Theta G^2 m_\pi^4 m_K^6}{24(4\pi)^5\, F^2 m_D^3} \tilde{I}(m_D/m_K) \tag{4.106}$$

---

[23]) Since the function $I(\varkappa)$ is approximately constant for the decays $\bar{D}^0 \to K^+ e\bar{\nu}_e$, $F^- \to \eta e\bar{\nu}_e$ and $\bar{K}^0 \to \pi^+ e\bar{\nu}_e$, the following approximate relations are valid

$$W_{(\bar{D}^0 \to K^+ e\bar{\nu}_e)} \approx \frac{m_D}{m_K} \left(\frac{m_K}{m_\pi}\right)^4 \cot^2 \Theta\, W_{(\bar{K}^0 \to \pi^+ e\bar{\nu}_e)}$$

$$W_{(F^- \to \eta e\bar{\nu}_e)} \approx \frac{2}{3} \frac{m_F}{m_K} \left(\frac{m_\eta}{m_\pi}\right)^4 \cot^2 \Theta\, W_{(\bar{K}^0 \to \pi^+ e\bar{\nu}_e)}.$$

where $\tilde{I}(y)$ is a phase-space integral defined by

$$\tilde{I}(y) = \int_1^{(y-1)^2} dx \sqrt{(y^2-1)^2 + x[x - 2(y^2+1)]} \left[x^3 - 8(x^2-1) - \frac{1}{x} + 12x \ln x\right].$$

We find

$$W_{(\bar{D}^0 \to K^0 \pi^+ e \bar{\nu}_e)} = 2.1 \cdot 10^6 \text{ s}^{-1}. \tag{4.107}$$

Analogously, we have

$$W_{(F^- \to K^+ K^- e \bar{\nu}_e)} = \frac{\cos^2 \Theta G^2 m_K^{10}}{24(4\pi)^5 F^2 m_F^3} \tilde{I}(m_F/m_K) \tag{4.108}$$

and

$$W_{(F^- \to K^+ K^- e \bar{\nu}_e)} = 1.1 \cdot 10^9 \text{ s}^{-1}. \tag{4.109}$$

V. Discussions and Outlook

In this review an extensive use has been made of the one-loop approximation which plays a crucial role in deriving the results of Sect. IV. Clearly, there immediately arises the question to what extent such an approach makes sense. Although we have no direct arguments in favour of the validity of the one-loop approximation, there exist a lot of hints which support the applicability of this approximation scheme. Let us recall the arguments proposed first by LEHMANN [*9b*]. According to the Adler-Weinberg theorem the first order terms of perturbation theory lead to exact results in the low-energy limit. Hence, at low energies higher order terms should give small corrections. Such a situation is wellknown from quantum electrodynamics where the first order terms of perturbation theory contain the main information at low energies, and this would even hold for large coupling constants.

One further reason for the usefullness of the one-loop approximation in chiral theories is as follows. There exist a lot of decay processes, e.g. $(\pi^0, \eta, \eta', K_L) \to 2\gamma$, $(\eta, \eta', K_L) \to \pi^+\pi^-\gamma$ etc., that proceed through anomalous loop diagrams [*10*]. These anomalous one-loop diagrams contain in principle the basic information on the processes under consideration, and higher order diagrams can be neglected.

As we have found for the chiral groups $SU(2) \times SU(2)$ and $SU(3) \times SU(3)$, the agreement between theory and experiment is in most cases satisfactory (there is the only exception of the decay $\eta \to \pi^0\gamma\gamma$). Thus, despite of the above mentioned lack of rigorous theoretical arguments in favour of the validity of our approximation scheme, we consider it to be justified mainly by showing its phenomenological successes.

Our approach to strong interactions which relies on nonpolynomial chiral Lagrangians is essentially a low-energy approach. Thus there naturally arises the question whether one can extend the chiral theories to intermediate and high energies or, alternatively, whether chiral theories may be considered to be limiting cases of other theories possessing suitable high energy properties.

Let us remind the reader that there exists nowadays a very fashionable theory of strong interactions, quantum chromodynamics (QCD), a gauge theory of colored quarks and gluons. By construction, this theory is very suited to describe the parton-like structure of hadrons reflecting itself in the scaling behaviour (and specific violations of it) of deep inelastic lepton-hadron processes. On the other hand, very little is known at present about the low-energy or large distance behaviour of this theory. This concerns such important questions as quark confinement (needed to explain the nonobservability of quarks) and symmetry properties of the vacuum as well as the calculability of the physical hadron spectrum (some phenomenological attempts in this direction have, however, been made on the basis of Schroedinger equations [*86*] and bag models [*87*]). Thus, a challenging task to be solved in the near future is to close the gap between "microscopic" strong interaction Lagrangians of the QCD-type expressed in terms of nonobservable fundamental particles and phenomenologically successful low-energy Lagrangians formulated in terms of observable hadrons[24]). In this connection we note that for two-dimensional QCD is has been shown that an effective meson Lagrangian can be deduced by passing over to new dynamical field variables [*89*]. Moreover, Polyakov [*90*] and Aref'eva [*91*] obtained a very interesting representation of four-dimensional QCD in terms of fields defined on a contour. It is impressive that in terms of these new contour variables the QCD-Lagrangian gets a form resembling nonpolynomial chiral Lagrangians.

Acknowledgements

We are very much indebted to our colleagues and friends of the Laboratory of Theoretical Physics for fruitful discussions and collaboration. One of the authors (D.E.) would also like to thank Profs. N. N. Bogolubov and V. A. Meshcheryakov for the permanent support rendered to him during his stay at the Joint Institute for Nuclear Research, Dubna.

VI. Appendices

Appendix A

Let us quote the explicit expressions for the 4×4 meson matrix

$$P = \sqrt{2} \sum_{i=1}^{15} V_i \Phi_i$$

and the tensor representation of the baryon wave functions. We have

$$P = \begin{pmatrix} \frac{\pi^0}{\sqrt{2}} + \frac{\eta_8}{\sqrt{6}} + \frac{\eta_{15}}{\sqrt{12}} & \pi^+ & K^+ & \bar{D}^0 \\ \pi^- & -\frac{\pi^0}{\sqrt{2}} + \frac{\eta_8}{\sqrt{6}} + \frac{\eta_{15}}{\sqrt{12}} & K^0 & D^- \\ K^- & \bar{K}^0 & -\frac{2}{\sqrt{6}}\eta_8 + \frac{\eta_{15}}{\sqrt{12}} & F^- \\ D^0 & D^+ & F^+ & -\frac{\sqrt{3}}{2}\eta_{15} \end{pmatrix} \tag{A.1}$$

where

$$\pi^\pm = \frac{1}{\sqrt{2}}(\Phi_1 \mp i\Phi_2), \qquad D^\pm = \frac{1}{\sqrt{2}}(\Phi_{11} \pm i\Phi_{12}), \qquad \bar{D}^0 = \frac{1}{\sqrt{2}}(\Phi_9 - i\Phi_{10}) \quad \text{etc.},$$

[24]) Recently some interesting papers have been appeared in this direction (see [*88*]).

and [32]

$$
\begin{aligned}
&B_1^{[12]} = \frac{A_2^+}{\sqrt{6}} - \frac{B_2^+}{\sqrt{2}}, && B_1^{[13]} = \frac{A_1^+}{\sqrt{6}} + \frac{B_3^+}{\sqrt{2}}, && B_1^{[14]} = \frac{\Lambda}{\sqrt{6}} + \frac{\Sigma^0}{\sqrt{2}} \\
&B_1^{[23]} = B_3^{++}, && B_1^{[24]} = \Sigma^+, && B_1^{[34]} = P \\
&B_2^{[12]} = \frac{A_2^0}{\sqrt{6}} - \frac{B_2^0}{\sqrt{2}}, && B_2^{[13]} = B_3^0, && B_2^{[14]} = \Sigma^- \\
&B_2^{[23]} = \frac{A_1^+}{\sqrt{6}} - \frac{B_3^+}{\sqrt{2}}, && B_2^{[24]} = \frac{\Lambda}{\sqrt{6}} - \frac{\Sigma^0}{\sqrt{2}}, && B_2^{[34]} = N \\
&B_3^{[12]} = -B_1^0, && B_3^{[13]} = \frac{A_2^0}{\sqrt{6}} + \frac{B_2^0}{\sqrt{2}}, && B_3^{[14]} = -\Xi^- \\
&B_3^{[23]} = -\left(\frac{A_2^+}{\sqrt{6}} + \frac{B_2^+}{\sqrt{2}}\right), && B_3^{[24]} = \Xi^0, && B_3^{[34]} = -\sqrt{\frac{2}{3}}\Lambda \\
&B_4^{[12]} = -C_1^+, && B_4^{[13]} = C_2^+, && B_4^{[14]} = -\sqrt{\frac{2}{3}}A_2^0 \\
&B_4^{[23]} = -C_2^{++}, && B_4^{[24]} = \sqrt{\frac{2}{3}}A_2^+, && B_4^{[34]} = \sqrt{\frac{2}{3}}A_1^+ \\
&B_i^{[ik]} = 0
\end{aligned}
\tag{A.2}
$$

Appendix B

For completeness, we include in this appendix the lepton sector of the Weinberg-Salam-model written in a compact 4-dimensional notation. Using the matrix representation (3.7) of the generators of the group $SU(2)_L \times U(1)$ the covariant derivatives of the lepton and Higgs multiplets l, Φ_L

$$
l = \begin{pmatrix} \nu_e \\ e \\ \mu \\ \nu_\mu \end{pmatrix}, \qquad \Phi_L \equiv \frac{1+\gamma_5}{2} \begin{pmatrix} \varphi_+ \\ \varphi_0 \\ \varphi_0 \\ \varphi_+ \end{pmatrix} \tag{B.1}
$$

may be written as

$$
D_\mu(l; \Phi_L) = (\partial_\mu + i\varkappa W_\mu \cdot \hat{C} + i\varrho X_\mu \hat{Y}/2)\ (l; \Phi_L) \tag{B.2}
$$

where Y_W takes the values -1 or $+1$ for l or Φ_L, respectively. The Lagrangian of the Weinberg-Salam-model then reads ($l_L = 1 + \gamma_5/2\ l$)

$$
\begin{aligned}
\mathcal{L}_{WS} = &-\frac{1}{4}(W^k_{\mu\nu})^2 - \frac{1}{4}(X_{\mu\nu})^2 + \frac{1}{4}\operatorname{Tr}(D_\mu\Phi_L)^+(D^\mu\Phi_L) \\
&- V(\Phi_L^+\Phi_L) + i\bar{l}\gamma_\mu D^\mu l - \frac{1}{2}\operatorname{Tr}(\bar{l}_L A \Phi_L + \Phi_L^+ \bar{A} l_L)
\end{aligned}
\tag{B.3}
$$

where $W^k_{\mu\nu}$, $X_{\mu\nu}$ are the usual covariant curls of the gauge bosons $W_\mu{}^k$, X_μ respectively. The Higgs potential may be cast into the form

$$V(\Phi_L{}^+\Phi_L) = \frac{\lambda}{8} \operatorname{Tr} \left(\Phi_L{}^+\Phi_L - \frac{1+\gamma_5}{2} v^2\right)^2. \tag{B.4}$$

The last term in eq. (B.3) generates the lepton masses. Using the expression

$$A = \begin{pmatrix} f_1 e_R I & 0 \\ 0 & f_2 \mu_R I \end{pmatrix}, \quad I = \begin{pmatrix} 1 & 0 \\ 0 & 1 \end{pmatrix}, \quad \left(e_R = \frac{1-\gamma_5}{2} e\right) \tag{B.5}$$

and

$$\langle\Phi_L\rangle_0 = \frac{1+\gamma_5}{2} \begin{pmatrix} 0 \\ v/\sqrt{2} \\ v/\sqrt{2} \\ 0 \end{pmatrix} \tag{B.6}$$

we get $m_e = f_1 v/\sqrt{2}$, $m_\mu = f_2 v/\sqrt{2}$. In addition, there arise the following vector boson masses

$$(M_W{}^2)_{\text{Higgs}} = \left(\frac{\varkappa v}{2}\right)^2; \qquad (M_Z{}^2)_{\text{Higgs}} = (\varkappa^2 + \varrho^2) \left(\frac{v}{2}\right)^2. \tag{B.7}$$

Let us consider the quadratic part of the Lagrangian (B.3) $\left(\varphi_0 = 1/\sqrt{2}\, v + (\sigma + i\chi)\right)$

$$\begin{aligned} \mathscr{L}_{\text{quad}} = {} & |\partial_\mu \varphi_+|^2 + \frac{1}{2} [(\partial_\mu \sigma)^2 + (\partial_\mu \chi)^2] + \overline{M}_W{}^2 \, |W_\mu{}^+|^2 \\ & + \frac{1}{2} \overline{M}_Z{}^2 Z_\mu{}^2 - iM_W[\partial^\mu \varphi_+ W_\mu{}^- - \partial^\mu \varphi_- W_\mu{}^+] - M_Z \partial^\mu \chi Z_\mu . \end{aligned} \tag{B.8}$$

In eq. (B.8) we have included the mass terms of the vector bosons arising from their interaction with the 0^--mesons. To get the vector boson propagators in the R-gauge we choose the gauge conditions

$$\begin{aligned} \mathscr{L}_g = {} & -\frac{1}{2\alpha} (\partial_\mu A^\mu)^2 - \beta \left|\partial_\mu W^{\mu+} + i \frac{M_W}{\beta} \varphi_+\right|^2 - \frac{1}{\beta} \left(\frac{f\varkappa}{\sqrt{2}}\right)^2 |\varphi_+|^2 \\ & - \frac{\gamma}{2} \left(\partial_\mu Z^\mu + \frac{M_Z}{\gamma} \chi\right)^2 - \frac{1}{4\gamma} \left(\frac{f\varkappa}{\cos \Theta_W}\right)^2 \chi^2. \end{aligned} \tag{B.9}$$

This yields the standard expressions

$$\begin{aligned} iD^W_{\mu\nu} &= \left(-g_{\mu\nu} + \left(1 - \frac{1}{\beta}\right) \frac{k_\mu k_\nu}{k^2 - \dfrac{\overline{M}_W{}^2}{\beta} + i\varepsilon}\right) \frac{i}{k^2 - \overline{M}_W{}^2 + i\varepsilon}, \\ iD^{\varphi_+} &= \frac{i}{k^2 - \dfrac{\overline{M}_W{}^2}{\beta} + i\varepsilon} \quad \text{etc.} \end{aligned} \tag{B.10}$$

As usual, the ghost pole at $k^2 = \overline{M}_W{}^2/\beta$ in the vector boson propagator cancels the pole in the Goldstone (φ_+)-propagator.

Appendix C

Let us indicate the formula for the expression $\mathscr{L}_{\rm bil}$ (cf. eq. 3.15) which is bilinear in the vector fields (the prime now means that we here included also bilinear terms arising from the currents)

$$\begin{aligned}\mathscr{L}'_{\rm bil} &= \left(\frac{F\varkappa}{\sqrt{2}}\right)^2 |W_\mu{}^+|^2 + \left(\frac{F\varkappa}{2\cos\Theta_W}\right)^2 Z_\mu{}^2 \\ &+ \left(\frac{\varkappa F}{4\cos\Theta_W}\right)^2 Z_\mu{}^2 \sin^2\Theta_W \left\{H\left(\frac{C_3}{2}\right) - 2\sin^2\Theta_W\langle[Q, e^{i2\xi\cdot A}]\,[Q, e^{-i2\xi\cdot A}]\rangle\right\} \\ &- e^2 A_\mu{}^2 \frac{F^2}{8}\langle[Q, e^{i2\xi\cdot A}]\,[Q, e^{-i2\xi\cdot A}]\rangle - e\left(\frac{\varkappa}{2\sqrt{2}}\right)\frac{F^2}{8}(A_\mu W^{\mu+}H(C_+) + \text{h.c.}) \\ &+ \left(\frac{\varkappa}{2\sqrt{2}}\right)\left(\frac{\varkappa}{2\cos\Theta_W}\right) 2\sin^2\Theta_W \frac{F^2}{8}(Z_\mu W^{\mu+}H(C_+) + \text{h.c.}) \\ &- e\left(\frac{\varkappa}{2\cos\Theta_W}\right)\frac{F^2}{8} Z_\mu A^\mu \left\{H\left(\frac{C_3}{2}\right) - 4\sin^2\Theta_W\langle[Q, e^{i2\xi\cdot A}]\,[Q, e^{-i2\xi\cdot A}]\rangle\right\}. \qquad \text{(C.1)}\end{aligned}$$

Here the function $H(X)$ describes the coupling of the vector bosons and pseudoscalar mesons. It reads

$$H(X) = \langle[Q, e^{i2\xi\cdot A}]\,([X, e^{-i2\xi\cdot A}] - \gamma_5\{X, e^{-i2\xi\cdot A}\}) + (\gamma_5, A_i) \to (-\gamma_5, -A_i)\rangle. \qquad \text{(C.2)}$$

As we observe from eq. (C.1) there appear mass terms for the W and Z bosons due to their interactions with the hadron sector. These mass terms must be added to the masses arising from the ordinary Higgs mechanism in the lepton sector of the Weinberg-Salam model (cf. eq. (3.48) and Appendix B). The remaining terms in $\mathscr{L}_{\rm bil}$ describe the interaction of the pseudoscalar 15-plet (π, k, η_8, F, D, η_{15}) with the gauge bosons. Our model contains the following 3-particle vertices.

$$AW^\pm\xi\text{-vertex:}\quad ie\left(\frac{\varkappa}{2\sqrt{2}}\right)\frac{F^2}{2}\langle C_+(\Theta)\,[Q, V_i]\rangle A_\mu W^{\mu\pm}\xi_i$$

$$ZW^\pm\xi\text{-vertex:}\quad -i\left(\frac{\varkappa}{2\sqrt{2}}\right)\left(\frac{\varkappa}{2\cos\Theta_W}\right) 2\sin^2\Theta_W \frac{F^2}{2}\langle C_+(\Theta)\,[Q, V_i]\rangle Z_\mu W^{\mu\pm}\xi_i.$$

The analogous expression for an $AZ\xi$-vertex vanishes because of $\langle C_2[Q, V_i]\rangle = 0$. Note that (C.1) contains no "seagull" terms for the charged vector bosons.

Appendix D

We collect here some useful formulas for the meson currents obtained by expanding them in powers of the ξ field. We restrict ourselves to zero order contributions in e and $\varkappa$. Using some simple algebra the meson currents may be put in the form

$$J^0_{\mu,m} = -i\,\frac{F^2}{2}\langle \tilde{P} e^{i2\xi\cdot A}\,\partial_\mu e^{-i2\xi\cdot A}\rangle, \qquad \text{(D.1)}$$

$\tilde{P}$ being the Cabibbo rotated projection operator $\tilde{P} = \{2\hat{C}_+(\Theta),\ 2\hat{C}_3 - 2\sin^2\Theta_W Q, Q\}$ introduced in Sect. III.2.1. Taking into account

$$e^{i2\xi\cdot A}\,\partial_\mu e^{-i2\xi\cdot A} = \frac{1}{F^2}\left(-iF\,\partial_\mu\Phi_k\gamma_5 + if_{ijk}\Phi_i\,\partial_\mu\Phi_j\right)\lambda_k + O(\Phi^3) \tag{D.2}$$

one further has

$$\begin{aligned} J^{0W}_{\mu,m} = {} & \cos\Theta\left\{-F\sqrt{2}\,\partial_\mu\pi^- - F\sqrt{2}\,\partial_\mu F^-\right. \\ & \left. + (f_{ij1} + if_{ij2})\,\Phi_i\,\partial_\mu\Phi_j + (f_{ij13} - if_{ij14})\,\Phi_i\,\partial_\mu\Phi_j + O(\Phi^3)\right\} \\ & + \sin\Theta\left\{-F\sqrt{2}\,\partial_\mu K^- + F\sqrt{2}\,\partial_\mu D^-\right. \\ & \left. + (f_{ij4} + if_{ij5})\,\Phi_i\,\partial_\mu\Phi_j - (f_{ij11} - if_{ij12})\,\Phi_i\,\partial_\mu\Phi_j + O(\Phi^3)\right\} \end{aligned} \tag{D.3}$$

$$\begin{aligned} J^{0Z}_{\mu,m} = {} & -F\,\partial_\mu\left(\pi^0 + \frac{1}{\sqrt{3}}\eta_8 - \sqrt{\frac{2}{3}}\,\eta_{15}\right) \\ & + (1 - 2\sin^2\Theta_W)\left(f_{ij3} + \frac{1}{\sqrt{3}}f_{ij8} - \sqrt{\frac{2}{3}}\,f_{ij15}\right)\Phi_i\,\partial_\mu\Phi_j + O(\Phi^3) \end{aligned} \tag{D.4}$$

$$J^{0A}_{\mu,m} = \left(f_{ij3} + \frac{1}{\sqrt{3}}f_{ij8} - \sqrt{\frac{2}{3}}\,f_{ij15}\right)\Phi_i\,\partial_\mu\Phi_j + O(\Phi^4). \tag{D.5}$$

VII. References

[1] For excellent books and review articles on chiral symmetries see: S. L. Adler, R. F. Dashen. "Current Algebras" (W. A. Benjamin, Inc. New York—Amsterdam, 1968); S. B. Treiman, R. Jackiw, D. J. Gross. Lectures on Current Algebra and its applications" (Princeton University Press, New Jersey, 1972); V. de Alfaro, S. Fubini, G. Furlan and C. Rossetti, "Currents in Hadron Physics" (North-Holland Pub. Company, Amsterdam—London 1973); H. Pagels, Phys. Rep. **16**, 221 (1975); E. Reya, Rev. mod. Phys. **46**, 545 (1974).

[2] M. K. Volkov, V. N. Pervushin, "Essentially nonlinear quantum theories, dynamical symmetries and meson physics" (Moscow, Atomizdat 1978; in Russian).

[3] J. Schwinger, a) Ann. Phys. **2**, 407 (1957); b) Phys. Letters **B 24**, 473 (1967).

[4] M. Gell-Mann, M. Levy, Nuovo Cimento **16**, 705 (1960).

[5] F. Gürsey, Nuovo Cimento **16**, 230 (1960); P. Chang, F. Gursey, Phys. Rev. **164**, 1752 (1967).

[6] S. Weinberg, Phys. Rev. Letters **18**, 188 (1967).

[7] S. Coleman, I. Wess, B. Zumino, Phys. Rev. **177**, 2239 (1969); ibid. 2247.

[8] D. V. Volkov, Particles and Nucleus **4**, 3 (1973).

[9a] A. A. Slavnov, L. D. Faddeev, Teor. Mat. Fiz. **8**, 297 (1971); A. A. Slavnov, Nuclear Phys. **B 31**, 301 (1971); b) H. Lehmann, Phys. Letters **41**, 529 (1972); H. Lehmann, H. Trute, Nuclear Phys. **B 52**, 280 (1973).

[10] S. L. Adler, W. Bardeen, Phys. Rev. **182**, 1517 (1969); W. Bardeen, Phys. Rev. **184**, 1848 (1969); M. V. Terentiev, Uspechi. Fiz. Nauk **112**, 37 (1974).

[11] M. Levy, Nuovo Cimento **52 A**, 23 (1967).

[12] B. Hu, Phys. Rev. **D 9**, 1825 (1974).

[13] W. A. Bardeen, B. W. Lee, Phys. Rev. **177**, 2389 (1969), see also the review by S. Gasiorowicz, D. A. Geffen, Rev. mod. Phys. **41**, 531 (1969).

[14] J. Cronin, Phys. Rev. **161**, 1483 (1967); J. Wess, B. Zumino, Phys. Rev. **163**, 1727 (1967); B. Lee, H. T. Nieh, Phys. Rev. **166**, 1507 (1968).

[15] S. Weinberg, Phys. Rev. **166**, 1568 (1968).

[16] J. GOLDSTONE, Nuovo Cimento **19**, 154 (1961); J. GOLDSTONE, A. SALAM, S. WEINBERG, Phys. Rev. **127**, 965 (1962).
[17] F. GÜRSEY, Acta Phys. Austriaca, Suppl. V, p. 185 (1968).
[18] V. I. OGIEVETSKY, Proceedings of Xth Winter School of Theoretical Physics, Karpacz, vol. **1**, p. 117 (1974).
[19] G. ECKER, J. HONERKAMP, Nuclear Phys. **B 62**, 509 (1973).
[20] M. K. VOLKOV, V. N. PERVUSHIN, Uspechi Fiz. Nauk **120**, 363 (1976).
[21] D. EBERT, M. K. VOLKOV, JINR Prepr. E2-11958 (1978) and Proceedings of the Internal. Seminar on High Energy Physics, Protvino, vol. **1**, 154 (1978).
[22] S. WEINBERG, Phys. Rev. Letters **19**, 1264 (1967); A. SALAM, Elementary Particle Physics, ed by N. Svartholm, Stockholm, p. 367 (1968).
[23] D. EBERT, Nuovo Cimento **54 A**, 399 (1979).
[24] M. K. VOLKOV, D. EBERT, Yadern Fiz. **29**, 523 (1979).
[25] M. K. VOLKOV, D. EBERT, Yadern Fiz. **31**, 520 (1980).
[26] M. GELL-MANN, R. J. OAKES, B. RENNER, Phys. Rev. **175**, 2195 (1968); R. J. OAKES, Phys. Letters **29 B**, 683 (1969).
[27] A. EBRAHIM, M. SERDAROGLU, Phys. Letters **48 B**, 338 (1974).
[28] M. K. VOLKOV, Particles and Nucleus **10**, 693 (1979).
[29] E. D. BLOOM, Proceedings of the Internat. Symp. on Lepton and Photon Interactions at High Energies, Fermilab, 1979. (Ed. T. B. W. Kirk, H. D. I. Abarbanel).
[30] A. KAZI et al., Acta Physica Austriaca **45**, 65 (1976); Z. MAKI et al., Preprint Kyoto University, RIFP-255 (1976).
[31] R. H. SOCOLOW, Phys. Rev. **137 B**, 1221 (1965).
[32] M. KOBAYASHI et al., Progr. theor. Phys. **47**, 982 (1972).
[33] T. G. TRIPPE et al., Phys. Letters **68 B**, 1 (1977).
[34] S. L. GLASHOW, J. ILIOUPOULOS, L. MAIANI, Phys. Rev. **D 2**, 1285 (1970).
[35] M. SINGER, Phys. Rev. **D 16**, 2304 (1977); J. L. KALINOVSKY, V. N. PERVUSHIN, JINR Prepr. P 2-11358 (1978).
[36] W. OCHS, Preprint Max-Planck-Institut, MPI-PAE/PTh 32/77 (1977) (Invited talk at the II. Intern. Conf. on Nucleon-Nucleon Interaction, Vancouver, B. C., Canada, June, 1977).
[37] M. K. VOLKOV, V. N. PERVUSHIN, Nuovo Cimento **A 27**, 277 (1975).
[38] P. S. ISAEV, Particles and Nucleus **4**, 731 (1973).
[39] M. K. VOLKOV, Comm. Math. Phys. **7**, 289 (1968); Ann. Phys. **49**, 202 (1968), Fortschr. Phys. **22**, 499 (1974).
[40] M. K. VOLKOV, V. N. PERVUSHIN, Yadern. Fiz. **20**, 762 (1974).
[41] J. P. BATON et al., Phys. Letters **B 33**, 525; 528 (1970).
[42] S. D. PROTOPOPESCU et al., L. A. L. preprint 787 (1972).
[43] M. K. VOLKOV, V. N. PERVUSHIN, Yadern Fiz. **19**, 652 (1974); Phys. Letters **51 B**, 356 (1974).
[44] S. B. GERASIMOV, Yadern Fiz. **29**, 513 (1979).
[45] E. B. DALLY, et al., Phys. Rev. Letters **39**, 1176 (1977).
[46] A. S. ESAULOV, J. J. TITOV, Yadern Fiz. **27**, 1602 (1978).
[47] A. QUENZER et al., Phys. Letters **B 76**, 512 (1978).
[48] S. F. BERESHNEV et al., Yadern Fiz. **18**, 102 (1973).
[49] G. ADYLOV et al., Phys. Letters **B 51**, 402 (1974).
[50] M. K. VOLKOV, V. N. PERVUSHIN, Yadern Fiz. **22**, 346 (1975); Phys. Letters **B 52**, 405 (1975).
[51] M. V. TERENTIEV, Yadern Fiz. **16**, 162 (1972).
[52] M. K. VOLKOV, D. EBERT, JINR-prepr. P2-80-471 (1980) (Yadern. Fiz. (1981), in print).
[53] V. N. PERVUSHIN, M. K. VOLKOV, Phys. Letters **B 58**, 177 (1975).
[54] N. M. KROLL, T. D. LEE, B. ZUMINO, Phys. Rev. **157**, 1376 (1967).
[55] E. N. TSYGANOV, Proceedings of the XIX International Conference on High Energy Physics, Tokyo, p. 315 (1978).
[56] W. R. MOLZON et al., Phys. Rev. Letters **41**, 1213 (1978).
[57] M. V. TERENTIEV, Yadern Fiz. **19**, 1298 (1974).
[58] G. BACKENSTOSS et al., Phys. Letters **B 43**, 431 (1973).
[59] M. K. VOLKOV, V. N. PERVUSHIN, Yadern Fiz. **22**, 366 (1975).
[60] Rev. mod. Phys. **48**, No 2 (1976).
[61] I. STEINBERGER, Phys. Rev. **76**, 1180 (1949).

[62] D. J. BARDIN, S. M. BILENKIY, Yadern Fiz. **16**, 557 (1972); D. I. BARDIN, E. A. IVANOV, Particles and Nucleus **7**, 726 (1976).
[63] A. STETZ et al., Nuclear Phys. **B 138**, 285 (1978).
[64] M. PRATAP, J. SMITH, Phys. Rev. **D 5**, 2020 (1972).
[65a] G. S. ABRAMS et al., Phys. Rev. Letters **43**, 777 (1979); b) D. M. BINNIE et al., Phys. Letters **B 83**, 141 (1979).
[66] W. D. APEL et al., Phys. Letters **B 83**, 131 (1979).
[67] N. ISGUR, Phys. Rev. **D 12**, 3770 (1975).
[68] A. T. FILIPPOV, Yadern Fiz. **29**, 1035 (1979).
[69] G. S. GOUNARIS, J. J. SAKURAI, Phys. Rev. Letters **21**, 244 (1968).
[70] M. S. CHANOWITZ, Phys. Rev. Letters **35**, 977 (1975).
[71] D. EBERT, M. K. VOLKOV, Yadern Fiz. **32**, 503 (1980).
[72] V. M. BUDNEV, A. E. KALOSHIN, Phys. Letters **B 86**, 351 (1979).
[73] D. EBERT, M. K. VOLKOV, Yadern Fiz. **30**, 1420 (1979).
[74] N. D. H. DASS, Phys. Rev. **D 7**, 1458 (1973); G. J. GOUNARIS, Phys. Rev. **D 2**, 2734 (1970); G. OPPO and S. ONEDA, Phys. Rev. **160**, 1397 (1967).
[75] D. I. KAZAKOV, V. N. PERVUSHIN, M. K. VOLKOV. Phys. Letters **B 64**, 201 (1976).
[76] H. BRAUN et al., Nuclear Phys. **B 89**, 210 (1975).
[77] D. EBERT, D. V. KREOPALOV, M. K. VOLKOV, Yadern Fiz. **29**, 1581 (1979).
[78] a) L. ROSSELET et al., Phys. Rev. **D 15**, 574 (1977); b) E. W. BEIER et al., Phys. Rev. Letters **30**, 399 (1973); c) P. BASILE et al., Phys. Letters **B 36**, 619 (1971).
[79] F. A. BERENDS, A. DONNACHIE, G. C. OADES, Phys. Rev. **171**, 1457 (1968); A. K. MOHANT, R. E. MARSHAK, Nuovo Cimento **A 52**, 965 (1967).
[80] a) M. V. TERENTIEV, V. V. SOLOVIEV, Yadern Fiz. **16**, 153 (1972); b) I. WESS, B. ZUMINO, Phys. Letters **B 37**, 95 (1971).
[81] M. K. VOLKOV, Yadern Fiz. **28**, 162 (1978).
[82] M. K. GAILLARD, B. W. LEE, Phys. Eev. **D 10**, 897 (1974).
[83] J. SAKURAI, Phys. Rev. **156**, 1508 (1967).
[84] A. S. CARROLL et al., Preprint BNL-26697, Upton 1979.
[85] M. K. GAILLARD, B. W. LEE, J. L. ROSNER, Rev. mod. Phys. **47**, 277 (1975).
[86] E. EICHTEN et al., Phys. Rev. Letters **34**, 369 (1975).
[87] P. N. BOGOLUBOV, Ann. Inst. Henri Poincare 8, 163 (1967); For further references see e.g. J. KUTI, Proceedings of the 1977 CERN-JINR School of Physics, Nafplion, Greece, 1977.
[88] C. ROSENZWEIG, J. SCHECHTER, G. TRAHERN, Phys. Rev. **D 21**, 3380 (1980); P. DI VECCHIA, F. NICODEMI, R. PETTORINI, G. VENEZIANO, CERN Preprint TH 2898 — (1980); R. ARNOWITT, P. NATH — North Eastern preprint NUB-2445 (1980).
[89] D. EBERT, V. N. PERVUSHIN, JINR Preprint E2-10731 (1977); Teor. Mat. Fiz. **36**, 313 (1978); D. EBERT, V. N. PERVUSHIN, H. REINHARDT, Particles and Nucleus **10**, 1114 (1979).
[90] M. A. POLYAKOV, Phys. Letters **B 82**, 274 (1979).
[91] I. YA. AREF'EVA, Letters Math. Phys. **3**, 270 (1979).

Albert Einstein in Berlin 1913–1933

(Studien zur Geschichte der AdW der DDR)

Teil II: Spezialinventar

Bearbeitet von einem Autorenkollektiv unter der Leitung von CHRISTA KIRSTEN und HANS-JÜRGEN TREDER

1979. 296 Seiten — gr. 8° — Leinen 38,— M

Bestell-Nr. 762 701 1 (2168/7)

Der zweite Band der Einstein-Dokumentation ist ein Spezialinventar, das die in den Archiven (und Bibliotheken) der DDR vorhandenen Einsteiniana, also Schriftstücke, die von Einstein selbst oder in seinem Auftrag verfaßt wurden, die an ihn gerichtet waren oder auf ihn Bezug nehmen, in Regestenform verzeichnet, und zwar für den Zeitraum seiner Mitgliedschaft in der Berliner Akademie der Wissenschaften (1913/14—1932/33).

Ferner enthält dieser Band eine Zusammenstellung von Protokollauszügen derjenigen Akademiesitzungen, an denen Einstein entweder selber teilnahm und mitwirkte oder in denen auf ihn Bezug genommen wurde. Abgerundet wird das Spezialinventar durch ein Verzeichnis der in den Akademieschriften gedruckten Arbeiten Einsteins sowie ein Verzeichnis seiner Berliner Patentschriften.

Bestellungen durch eine Buchhandlung erbeten

AKADEMIE-VERLAG

DDR - 1080 Berlin, Leipziger Str. 3—4

INHALT

Die „FORTSCHRITTE DER PHYSIK“ dienen der Publikation zusammenfassender Berichte über aktuelle Teilgebiete der Physik. Neben Originalberichten in deutscher oder englischer Sprache werden deutsche Übersetzungen wichtiger fremdsprachiger Berichte abgedruckt.

Aus dem Inhalt der folgenden Hefte:

SCHLICKEISER, R.: Gamma Ray Astronomy and the Origin of Cosmic Rays

MACK, G.: Physical Principles, Geometrical Aspects, and Locality Properties of Gauge Field Theories

KLEINERT, H., MAKI, K.: Lattice Textures in Cholesteric Liquid Crystals

BUCHBINDER, I. L., FRADKIN, E. S., GITMAN, D. M.: Quantum Electrodynamics in Curved Space-Time

GALIĆ, H., GUBERINA, B., TADIĆ, D.: Quantum Chromodynamics and Parity-Violating Nucleon-Nucleon-Pion Coupling

Manuskripte werden nach Vereinbarung mit dem Autor von der Schriftleitung entgegengenommen.

www.ingramcontent.com/pod-product-compliance
Lightning Source LLC
LaVergne TN
LVHW042251300926
849161LV00003B/549